SECOND TO NONE

SECOND TO NONE

EDITED BY
JOHN CULLER AND CHUCK WECHSLER

LiveOak Press, Inc.
CAMDEN, SOUTH CAROLINA

Copyright© 1988 by
LiveOak Press, Inc.
Camden, South Carolina
ISBN 0-929822-02-1

FOREWORD

It started in the last days of 1982 with a phone call from Art Carter. He asked if I'd be interested in writing a story about Dan Lefever for a young magazine called *Sporting Classics*. Indeed I was.

The same question came around again a year later, this time from then-managing editor Joanna Craig, who wanted a piece on Charles Parker and his splendid guns.

A few months after that, it was John Culler on the phone, with a question that ultimately would prompt a major shift in the direction of my life: Would I accept an invitation to become firearms editor at *Sporting Classics?* I gave John two answers, one short, one long. Both boiled down to "yes."

In the long version, I told John that I was flattered by the invitation, that I considered it a compliment to my work as a writer, because the magazine he'd dreamed up fit almost perfectly with my own idea of what a sporting magazine should be. In the four years since, I haven't changed my mind about any of that.

During those years, the column has given me opportunities to write on subjects that I might otherwise have been forced to pass by, because the gun-writing market is a fairly narrow one that isn't always sympathetic to a researcher who loves exploring the quirky back-alleys of history.

Even more important, writing feature stories for *Sporting Classics* has offered a chance to approach firearms history systematically and in depth, which to my mind is the only approach that ever will fully elucidate the old, complex, and fascinating love affair between man and the sporting gun. In a world where things built by hand are growing increasingly scarce, it's easy to forget that the gun is an artifact in the truest sense — that, historically, the gun is as much a product and a touchstone of human emotion as it is a consequence of inventive genius.

That view of things is likely to get short shrift in the average outdoor magazine. But *Sporting Classics* isn't the average outdoor magazine. For one thing, it celebrates quality rather than quantity of sporting experience. For another, it offers variety in subject matter and points-of-view. Some general-interest outdoor magazines simply refuse to publish a gun piece by anyone other than their gun editors — a nice niche for the gun editors, perhaps, but less than ideal for readers. *Sporting Classics*, on the other hand, speaks in a multitude of voices, all of them worth listening to.

This collection of gun stories finds those voices brought together for the first time. With one exception, the stories are reprinted directly from the magazine, with typos and other minor flaws repaired. The exception is my story on Ansley Fox and his guns. Since that piece appeared in the fall of 1985, I have continued my research with an eye toward a book on Fox. Now, the book is well under way, and the editors at *Sporting Classics* were kind enough to let me revise the story somewhat for this collection, adding new material and clarifying some details.

The topics and authors presented here cover a broad range of interest and approach. What we share is a genuine love for the gun, for the sport that has shaped its evolution, and for the working of the human spirit that sets the sporting gun apart from all the rest.

MICHAEL McINTOSH

CONTENTS

ITALIAN RENAISSANCE IN STEEL
by Tony Atwill — 10

L.C. SMITH — THE YANKEE SIDELOCK
by David E. Petzal — 20

THE HENRY — GRANDADDY OF THE LEVER ACTIONS
by Rick Hacker — 28

FORGOTTEN CLASSICS
by Michael McIntosh — 36

MYSTIQUE OF THE MODEL 70
by David E. Petzal — 42

SECOND TO NONE
by Michael McIntosh — 48

WINCHESTER'S MAGIC DOUBLE
by Jerry Warrington — 55

ANSLEY H. FOX — THE MAN, THE LEGACY
by Michael McIntosh — 62

GRANDAD'S GUN
by Michael McIntosh — 72

BROWNING'S LAST CREDENTIAL
by Ted Sefing — 78

AMERICAN CLASSIC RIFLES — THE WORLD'S FINEST
by Art Carter — 88

DOUBLE VISIONS
by Michael McIntosh — 95

THE SURPRISING 28 — SOMETIMES A GREAT NOTION
by Michael McIntosh — 102

COMMEMORATIVE WINCHESTERS
by Rick Hacker — 109

THE CLASSIC TRAP GUNS
by Michael McIntosh _____ 116

MODEL 12 — THE PERFECT REPEATER
by David E. Petzal _____ 124

PATRIARCH OF THE GUN SHOPS
by Michael McIntosh _____ 130

BERETTA — FIVE CENTURIES OF GUNMAKING
by Michael McIntosh _____ 140

SPANISH TREASURE
by Terry Wieland _____ 148

THE HARD TIMES WONDER
by Michael McIntosh _____ 156

MY PURDEY — MUCH MORE THAN A GUN
by George Bird Evans _____ 164

DAD'S PUMP GUN
by Michael McIntosh _____ 172

ITHACA DOUBLES — OVERLOOKED CLASSICS
by Paul Rundell _____ 178

CANVAS OF STEEL
by David E. Petzal _____ 186

THE LEGEND OF UNCLE DAN
by Michael McIntosh _____ 194

SAVAGE MODEL 99
by Pete Laurie _____ 204

CHARLIE PARKER'S SHOTGUN
by Michael McIntosh _____ 210

CLASSIC QUARTET
by Rick Hacker _____ 218

SECOND TO NONE

Italian RENAISSANCE In Steel

The fantastic shotguns of Abbiatico & Salvinelli may be the most beautiful ever made. Today, fourteen skilled craftsmen turn out only fifty guns a year.

TONY ATWILL

It's fitting that the Roman goddess of the hunt, Diana, should appear on A&S guns. On the left sidelock of this double, another work of Fracassi, Diana and her nymphs are surprised in their bath by a mortal, the hunter Actaeon, who, for his indiscretion, is transformed by the goddess into a stag. On the bottom of the action is the shepherd Endymion, who was so handsome, according to mythology, that the moon, Selene, fell in love with him and lulled him into an eternal slumber so she might steal a kiss.

There is an old saw about engraving that goes something like this: "One of the nicest aspects of owning a well-engraved gun is that when the birds aren't flying, you still have something to look at." Like most old saws, this one holds true for the middle ground and falls apart in the extremes. If the engraving is truly bad, if it resembles chicken scratches in the sand when you look at it closely, you're apt to pick up, go home, and pawn the gun. If it is superb, on the other hand, you're likely to keep on staring as the birds fly by.

No doubt a lot of birds have flown over guns created these last few years by Abbiatico & Salvinelli of Gardone, Val Trompia, Italy. A few birds even may have landed to perch on the gunner's shoulder for a peek themselves, for the quality of gun and engraving surpasses anything ever done. *Anything.*

That may be hard to believe, but it is true, as the accompanying pictures testify. The reputation of English guns is so great that few of us dare think anything could surpass them in quality. They are inviolable. Furthermore, things Italian — other than Ferraris — often draw titters. Beautiful design, yes. Quality and workmanship, hardly. Even Italians will admit that reputation is often just that. Mario Abbiatico, the co-owner of Abbiatico & Salvinelli, commenting on the work of his gunsmiths and engravers, notes, ". . . it shows that the sense of beauty, love of one's work, art, and creative capacity is not altogether lost to the Italian people."

Hardly. The Italian Renaissance is about to be replayed, this time in steel.

"The finest double shotguns in the world and

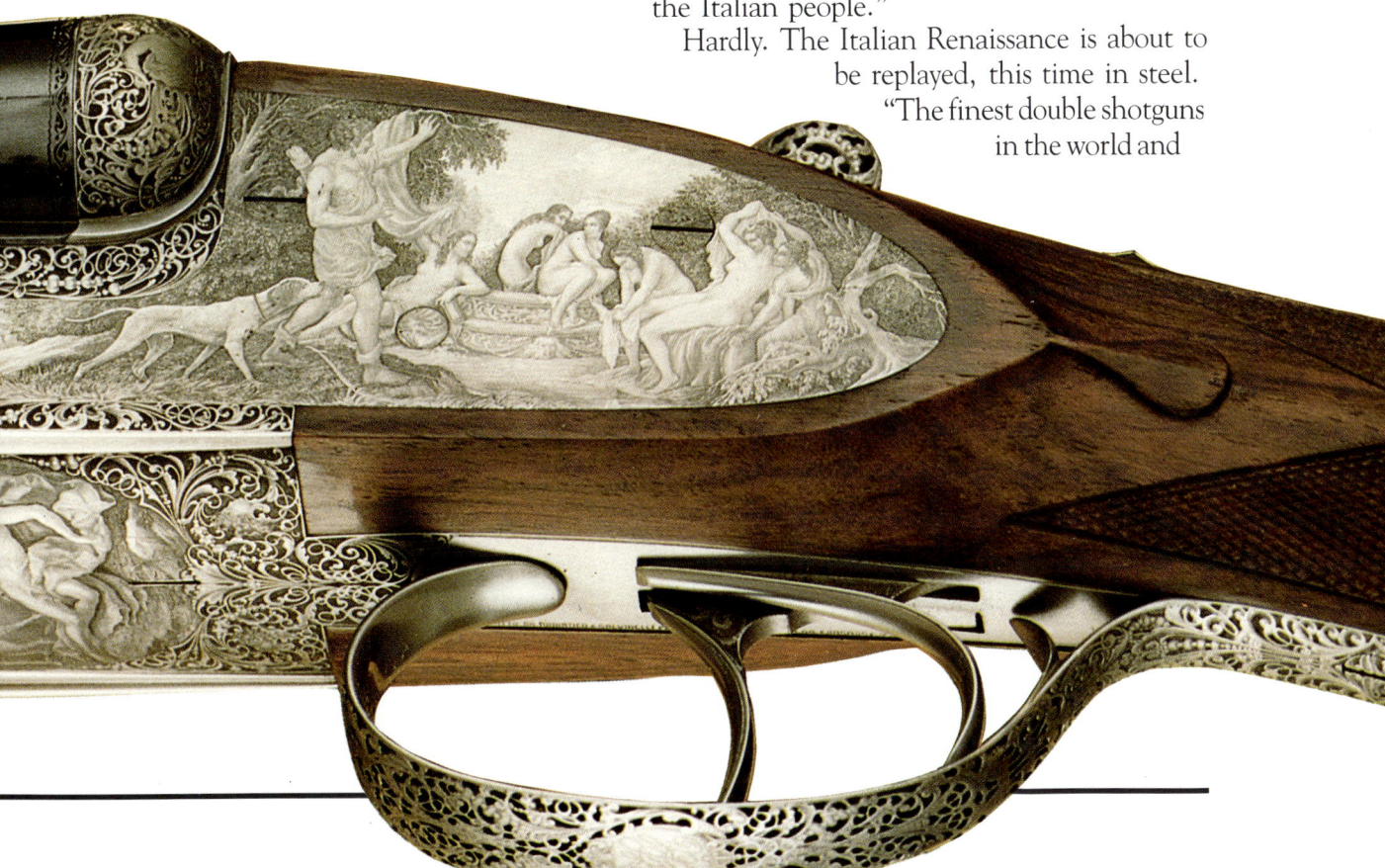

11

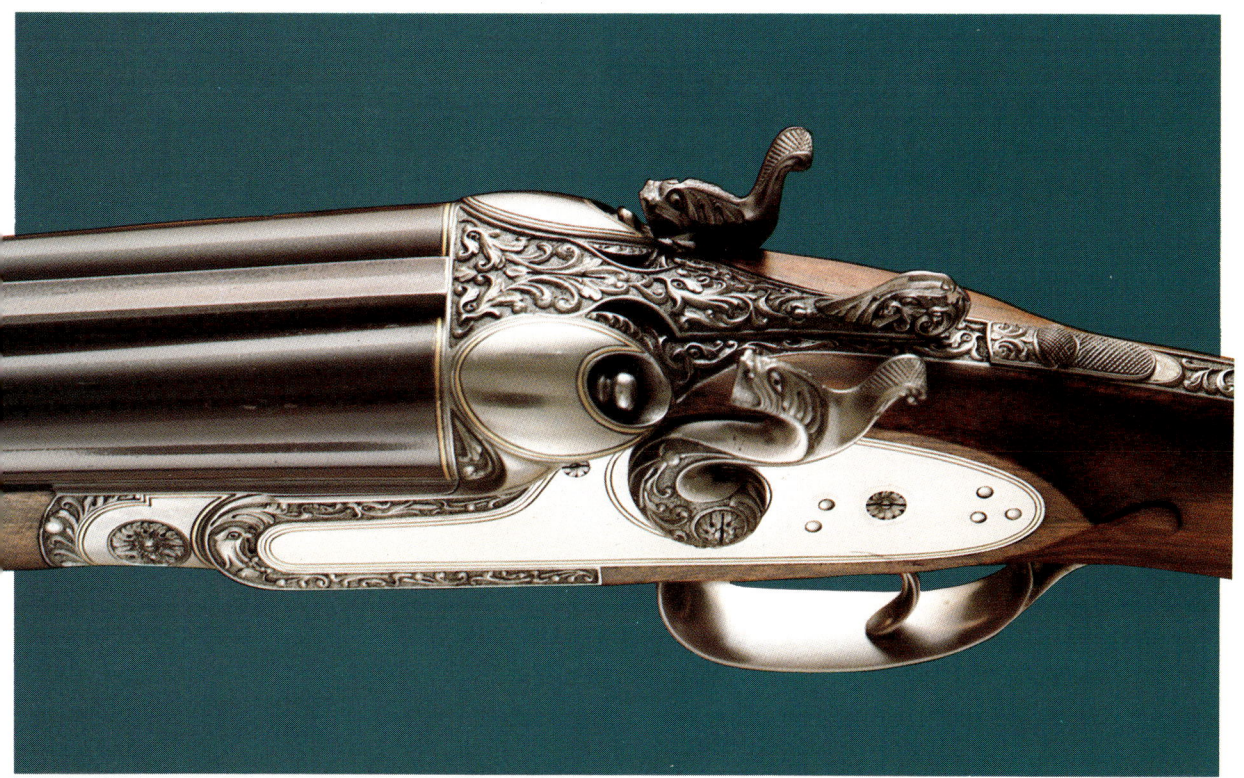

the best engraving are yet to be made," notes doubles collector Joseph Bojalad of Pittsburgh. "Tomorrow they [Abbiatico & Salvinelli] will make a better gun than they made today. The golden era that died in England in the thirties and forties, Italy has yet to see. They are the best in the world — in gun making and engraving — and they are going to get better." Bojalad speaks with some authority. He has a remarkable collection of fine doubles — the full spectrum of English and European guns — but he specializes in doubles by Abbiatico & Salvinelli. He has so many, in fact, that he may well be considered the firm's American patron, and indeed he is their American representative, only because of his love for the guns and his deep friendship with the two owners.

Left, the exceptional shading and modeling of this banknote engraving of two setters by Firmo Fracassi and the ornamental border show the delicacy of the bulino technique.

Exposed, automatic cocking hammers are a signature of Abbiatico & Salvinelli. The bottom gun is engraved with fine English scroll; this quality of work is representative of a "basic" A&S gun.

The top gun is decorated with gold inlay and deep chasing, a deep relief technique which is accented here with bright metal sideplates and breech.

Mario Abbiatico and Remo Salvinelli founded their company in 1967 under the name of Famars. Both had experience in the gunmaking trade; their northern Italian home of Gardone, Val Trompia is the hub of Italian gunmaking, a community of many talented artisans supplying parts and services to numerous large and small gunmakers in the area.

For ten years Famars made various grades of guns, from serviceable $200 doubles to masterpieces, but three years ago Abbiatico and Salvinelli decided to concentrate on only the best, guns that would bear their names alone. Today, 14 craftsmen make some 50 such guns each year. Everything is done in the A & S factory. About half of the production finds its way to the United States. All but half-a-dozen of those guns are built on commission. Production time runs about a year, sometimes less, and all work is taken with a guaranteed price, which starts around $8,000 for a sidelock 12 gauge, fully equipped with accessories and case and decorated with scrolls and rosettes, and goes up — way up — from there.

Abbiatico & Salvinelli offer three distinct models: side-by-sides with sidelocks in four gauges; the same gun with external, self-cocking hammers; and sidelocked over-and-unders in .410, 28 and 20 gauge. At times, of course, they will stray away from those styles: they are just finishing the last three guns in a limited production of 15 four-barreled 28 gauge shotguns (side-by-side-by-side-by-side?) with two internal hammers and two external hammers.

The quality of all their guns is superb. Some are made without visible pins or screws on the sideplates to give the engraver a flawless work surface. The wood-to-metal fit is exceptional, and the workmanship within — locks, ejectors and such — is mirror bright and smooth.

Mario Abbiatico, the co-owner of Abbiatico & Salvinelli, grew up in a family of gunmakers in Gardone, Val Trompia, Italy.

Left, top and bottom, Napoleon's campaigns inspired the engraving on this fine gun.

Furthermore, there is no question that A & S shotguns are made to be used, for that is a dictum among Italian gun makers: *Guns are for shooting.*

It is the engraving, however, that jumps out at you, engraving so good, so unique, that photographs hardly do it justice and words can but hint at its quality. The engravers who work for Abbiatico & Salvinelli are masters of numerous techniques. Their scrolls and rosettes — the English engraving most acceptable to the purist — are fine and precise and always enhance the basic lines of the gun rather than compete for attention. The inlaying is masterful, as is their chasing and filigree, but it is the technique known as *bulino* that is most unusual — and most grand.

To celebrate America's bicentennial, A&S engraved four guns — 12, 20, 28 and .410 — with scenes from the Revolution. The signing of the Declaration of Independence — a scene that included 42 people — adorns the bottom of the action of each gun; eight separate battle scenes decorate the eight sideplates.

Most engraving as we know it — border outlines, scrolls, rosettes around screws and pins, even scenes — is done with an engraving tool — a small chisel-like implement called a graver — which is tapped through the metal by a light, broadfaced hammer called a chasing hammer. The resulting cuts are sure and fluid in the hands of a master, but obviously a certain amount of delicacy is lost when the power to move the graver through the steel is transmitted by another tool.

The bulino technique, however, does not employ a chasing hammer. Rather, the engraver holds his graver in his hands and manipulates it through the steel with hand-power alone. The results are delicate cuts and dots that resemble the fine detail in an engraved printing plate, a banknote plate, for example, hence this style is often called banknote engraving.

In the hands of a master, bulino yields infinite detail and shading, which lend themselves not to decorative engravings but to ornamental work — scenes of game and dogs, animals, mythological events, primitive man, battles, historical occasions, and whatever else may inspire the engraver or his customer.

This style, on shotguns, is unusual to the American eye. Once the viewer can see past tradition, however he will recognize the far-reaching dimension this subtle technique brings to the engraver's art.

The bulino technique only succeeds when done extremely well. Poor technique on a faulted subject results in a caricature — Monopoly money instead of banknotes. But the engravers at A & S produce no caricatures. They are redefining the limits of the engraver's craft.

Everything you see in these photographs, every nuance of shading, is the result of a cut in steel. There is no blueing to the background,

Two sidelock over and under guns: the top a 12 gauge, the bottom a .410. The sidelock superimpose in .410 and 28 gauge is one of three standard A&S models; the price for the gun starts at $11,000 and goes up, depending on engraving. A 20 gauge will be available in this model towards the end of the year. This .410 weighs only 5.07 pounds. Engraving by Fausto Galeazzi.

This 28 gauge side-by-side, unfinished after four years of work, is perhaps the epitome of A&S work. To give the engraver, Firmo Fracassi, a flawless surface on which to depict these birds-of-prey, A&S built the action with no visible pins or screws.

no inking. The metal is bare, and prints can be pulled from the sidelocks. Every scene is but a play of light on steel.

But do not think that Abbiatico & Salvinelli guns are only showpieces. They are built, first of all, as guns to be fired, and the firm puts just as much effort in producing perfect lines in the overall design of the gun as it does in the engraving. "A fine gun is beautiful and valid in itself, whether it is engraved or not," says Mario Abbiatico. "The engraving is a part, sometimes modest and at other times important, in the construction of a firearm. A gun is valid for its ballistic qualities (the gun was created and constructed to shoot), for its mechanics, for its lines. The engraving can be left out without rendering the gun less valuable or less fine. If it is true, and it is, that engraving is used to embellish the gun, then one has to be careful not to spoil the lines, the proportions, the personality and the elegance. With the addition of too many intrusions (inscriptions, shields, engraving on the barrel, etc.), especially if these have not been chosen in relation to the rest, one runs the risk of spoiling the final result, not only of the engraving but of the gun itself.

"I believe there is a rule which one should keep to establish if there is something which interferes with the balance and the character of the gun. On examining it, even for the first time, the eye should pass over it without stopping or being disturbed or attracted to any particular detail. It is, in fact, the same rule which is used when examining any gun, be it engraved or not."

The eye does not stop when it admires a gun by Abbiatico & Salvinelli, nor can the mind help but marvel that the very best is yet to come.•

L.C. SMITH
The Yankee Sidelock

In the Golden Age of Shotgunning, the L.C. Smith was America's only sidelock double. And while she was not a perfect gun, Sweet Elsie was graceful and elegant enough to attract more than her share of admirers.

DAVID E. PETZAL

This 2-E Grade L.C. Smith produced in 1905 is typical of the pre-1913 Elsies.

To understand the guns, you must first understand something of the times that produced them. The half-century from 1890 to 1940 took America from the last of the Indian wars to the eve of Pearl Harbor, and it also encompassed what is now known as the Golden Age of Shotgunning.

This was an era when America was still rural. Superhighways and shopping malls were undreamed of. Farms were still small, and farmers saw no sin in leaving the occasional patch of cover. If the ground was not harvested clean of every solitary grain of wheat, it would not adversely affect corporate profits. The environment, the single most important element in determining game populations, was largely intact.

There were a lot of birds to be hunted, and a lot of time in which to hunt them. If you examine photos of the era, you will see men posing with days' bags that seem incredible by today's standards. Standing in their dark pants and vests and open white shirts, with their long-barreled shotguns at port arms or parade rest, they are the lucky ones.

The gun. In those days, there was only one type that really counted — the side-by-side. There were single-shots and pumps and recoil-operated autos, but they were ungainly objects, suitable for meat hunters, lacking in form and grace and style.

The double was the gun that could make a boy dream or light a grown man's eyes. And the most famous doubles came not from England, where names like Rigby and Westley Richards stirred men's souls, but from places like Ithaca, New York, and Meriden, Connecticut. The names of those guns formed a litany for the times: Parker, L. C. Smith, Ithaca, Lefever, and A. H. Fox.

Although each make had its adherents, the one that emerged as the premier of the Golden Age was the Parker. But the others were great in their time, and they bear a close look by collectors — especially America's only sidelock double, Sweet Elsie, the L. C. Smith.

Somewhere, there may be a line of firearms with a more unlikely pedigree than that of L. C. Smith, but it would be hard to find. "American Arms and Armsmakers," compiled in 1938 says of the firm:

"Established at Syracuse, N.Y., by Lyman Cornelius Smith in 1877. Smith sold his interests in 1890. During his term as head of the business he gave employment to an average of 176 workmen and produced more than 30,000 shotguns. The Smith line is now made by the Hunter Arms Co., Fulton, N.Y."

This is like saying that "Hamlet" is a play about a man who is very irritated at his uncle. The full story is much richer, and begins with a gunmaker named William Henry Baker who designed a drilling. This three-barreled rifle/shotgun was patented in 1875, and had been made by Baker in comparatively small numbers. By 1877, he had moved to Syracuse, New York, where he entered into a partnership with Lyman (L. C.) Smith and his brother, Leroy Smith. Under the name W. H. Baker and Company, the firm offered both the drilling and a double shotgun.

The company prospered, but by the end of the decade, both Leroy Smith and Baker had sold out to Lyman Smith, who continued to manufacture the Baker gun under the name L. C. Smith & Co.

The late nineteenth century was a time of great change and innovation in firearms, and L. C. Smith was shrewd enough to know that his guns were becoming dated. Salvation arrived in the form of one Alexander T. Brown, who had been with Smith since 1878 and who, in February 1883, applied for a patent on what he called a "Smith top-action, double-cross-bolted breech-loading gun."

This patent marked the beginning of the true L. C. Smith, for it was Brown's sidelock design that set the form for Elsies for the next sixty-plus years, and the first of the new guns, when they appeared in 1884, were stamped "L. C. Smith."

To appreciate what Brown wrought, and thus appreciate the Smith, a brief lecture on shotguns must be absorbed. There are two basic forms of modern double shotgun action: sidelock and boxlock. The sidelock carries the firing mechanism attached to two long sideplates that extend rearward from the breech. There are two

advantages in the sidelock system. Because the flat springs are longer than those in a boxlock, and therefore more lightly stressed, it is easier to open and cock the gun, and it is easier to obtain a superior trigger pull. The other advantage is artistic. The extended sideplate continues the line of the breech, giving a more graceful profile, and offers an engraver a much larger surface to work on.

Traditionally, the finest doubles have been sidelocks, but they have one major disadvantage. Unless the locks are inletted very carefully, the stock will be weak, as it lacks the "chest," or broad shelf of wood, that a boxlock incorporates where the action abuts the stock. In other words, there is no large bearing surface to absorb the force of recoil.

The boxlock is by far the more common, and was shared by the other great makes of the time. It incorporates the firing mechanism in the lockplate, and is, in its modern form, quite as good mechanically as the sidelock.

Brown introduced two other innovations. The first was a rotary locking bolt that was located at the top of the standing breech and which pivoted through a rib extension. The rib extension had a rectangular hole cut out of it, and the bolt fitted through this hole and over a lip at the end of the extension.

The purpose of this is to resist the bane of all double guns — shooting loose, or the development of play between the barrels and the breech, which can render a gun unshootable. Every time a double shotgun is fired, the muzzle is wrenched up by the force of the burning powder, and this constant twisting upward is what works a gun loose.

The trick in preventing this is to lock the barrels as solidly to the breech as possible, and the Smith system was unique. The Smith locking bolt, rather than passing laterally through the rib extension, passed up, through, and down over it, pulling the rear end of the barrels down with considerable force. If you pick up a Smith, remove the fore-end, and try to wiggle the barrels, they'll feel absolutely tight. But, if you swing the top lever to the side, removing the locking bolt from the rib extension, and then wiggle, you'll find that there is indeed play between barrels and breech.

Smith made a great deal of this locking system, and in the company's 1918 brochure, we are advised on every other page that "Smith guns never shoot loose." They do, just as any gun can, given sufficient time and wear.

Brown's other innovation came in 1886, with the patent for the L. C. Smith hammerless gun, and consisted of rotation cocking bars. When other shotguns used (and use) twin spurs or tabs that are pushed straight down to cock the action, Brown employed two cranklike levers that engaged recesses in the fore-end iron. When a Smith is broken, the levers are rotated downward about 40 degrees, and cams inside the receiver cock the hammers. It is indeed unconventional, and a good system.

The L. C. Smith gun was a success. It was made in both hammer and hammerless models, in a range of six different grades, numbered 2 to 7. They were available in 10 and 12 gauges only and were fitted with twist, or damascus, barrels. (The hammer gun remained in the line until 1934, a tribute to its popularity in the face of more modern designs.)

But in 1889, the L. C. Smith Gun Company went up for sale, due to — of all things — the typewriter. Alexander Brown had become intrigued with the new device and developed a model that was superior to the ones then in existence. He persuaded Smith that the future lay

A real Northwoods veteran is the 3-E grade L.C. Smith, made about 1906.

in upper and lower case type, rather than side-by-side barrels, and history proved him correct. The Smith Premier Typewriter Company eventually became Smith-Corona, and finally Smith-Corona-Marchant. L. C. Smith died in 1910, a rich man.

His gun business was purchased by the Hunter Arms Company of Fulton, New York, and in 1890, the first guns emerged from the Hunter plant, marked Hunter Arms Company, Fulton, N.Y. The basic design that Alexander Brown created was never changed. Hunter offered some new grades, and introduced modern steel barrels. The one significant development credited to Hunter Arms appeared in 1904, and bore the name The Hunter One Trigger. It was, as you might surmise, a single trigger. The One Trigger was selective, and although complex, it worked well.

In 1913, Hunter revised its entire line, and post 1913 Smiths can be identified by the grade of the gun stamped on the right-hand barrel near the breech. A dizzying array of models and options was offered. The 1918 catalogue lists no fewer than thirty-six variations, ranging from the economy model Fulton boxlock at $32.50 to the Delux, at $1,000 even. (The Fulton was something of an aberration, and is not considered an L. C. Smith shotgun.) Each successive grade offered more elaborate

The late 19th century was a time of great innovation in firearms, and L.C. Smith was shrewd enough to know that his guns were becoming dated. But salvation arrived in the form of Alexander Brown.

A 2-E Grade and pages from the 1905 L.C. Smith catalogue.

engraving, fancier wood and checkering, and refinements such as the One Trigger and automatic ejectors. Smiths were made in every gauge from 12 through .410, with the probable exception of 28. No one knows why, but apparently none were made in this bore size. Each gauge had its own frame size, and the smaller gauges are considerably trimmer than the larger bores. If I recall correctly, Larry Koller, the late gun writer, stocked a 20-gauge Elsie for his wife, and the gun weighed 4½ pounds.

The Depression and World War II spelled the end of Hunter Arms, as they did to the other great makers of doubles, and in 1945, Hunter Arms was sold to Marlin Firearms in New Haven, Connecticut. Marlin cut down drastically on the number of models produced, and the Smiths made in New Haven are marked "The L. C. Smith Gun Company, Inc."

Six years later, the end seemed to have come. The public's taste had turned to repeaters, and those who wanted two barrels preferred the over/under. It appeared that Sweet Elsie was gone forever.

But not so. In early 1967, Marlin re-issued the L. C. Smith in two grades, Field and Deluxe. These were plain guns with case-hardened locks, 28-inch barrels, and vent ribs. They were made in 12 gauge only. The new Elsies improved significantly on the old models. The inletting around the sidelocks was reinforced with fiberglass and the locks were positioned with jackscrews, which further took the strain off the wood. The stock dimensions were more modern, that is, they had less drop. The Field Grade, when introduced, sold for only $200, but it was not a success. Between 1968 and 1971, only 2,539 of the new Elsies were made, and then the curtain was drawn for good.

Collectors must view the Elsie with a critical eye. It was a good gun, but flawed, and can present problems to the person who wants to shoot it. The main problem is the sidelocks, and their tendency to crack the stock under use, and thus interfere with the trigger mechanism. One authority with whom I talked claims this is a problem only with the lower-grade guns, and that it is rarely seen in the higher grades where fitting of wood to metal was more careful. Another says that all Smiths cracked sooner or later.

If you are thinking of investing in one, it is imperative that you have the sideplates removed and the wood examined for damage. At the same time, have the gun checked for tightness and tinkering. Some Elsies were handed to the local blacksmith for repair, and these gents, having never seen a sidelock before, did some horrendous things to the mechanisms. These "repairs" can come back to haunt you.

From a shooter's viewpoint, the Smith presents a problem in that the pre-1967 guns are stocked with far more drop than a modern shotgun, and require that you shoot with your head far more erect than is standard nowadays. If this proves difficult, as it probably will, you can have the stock bent upward, or you can have a new stock made entirely. If you choose the former, the stock may break. If you decide to have a new stock made, have it done by someone who

Always buy from a recognized firearms dealer. The guy who offers you a Hunter Arms 20-gauge Specialty Grade for $350 may indeed have had to raise money for his aunt's prefontal lobotomy — or he may have just stolen it.

has already worked with sidelock guns. Do not get rid of the old stock. Keep it against the day you may want to sell the gun to a collector.

On today's market, it is the Parker and the Best-Grade English guns that are commanding high prices. Smiths are quite reasonable by comparison. A Hunter Arms Field Grade gun can run anywhere from $300 in 12 gauge to $650 in .410. The higher grades — Olympic, Specialty, Crown, and the others — can cost substantially more. The smaller gauges are the more desirable, and Ideal Grade in good condition, bored for 20-gauge shells, can bring around $4,500 today. When you compare this to the $30,000 to $60,000 that a really top-flight small-gauge Parker commands, the Smith's is indeed a moderate price.

Consider also that the market for collectible guns is subject to whim and fad. The slightly demented prices now paid for other guns may quite possibly push collectors to the Smith, and thus drive up its price. Twenty-five hundred dollars may be looked on as the ground floor in a few years.

You must buy from a recognized dealer. The guy who offers you the 20 gauge Specialty Grade for $350 may indeed have had to raise money fast for his aunt's prefontal lobotomy — or he may have just stolen it.

Be acutely aware also that any alteration, repair, or change to an original gun detracts from its value as a collector's piece. And that means *any* alteration.

Since there were so many different grades and models offered over the years, you are advised — yea, implored — to do some reading on the subject and know just which gun is which. Toward this end, I highly recommend *L. C. Smith Shotguns,* by Lt. Col. William S. Brophy, published by Beinfeld Publishing Inc., North Hollywood, California. It is profusely illustrated and carries a wealth of detail. For a lively history of the L. C. Smith I recommend *The Best Shotguns Ever Made in America,* by Michael McIntosh, published by Charles Scribner's Sons, New York, N.Y. It deals — in eminently readable terms — with Parker, L. C. Smith, Ithaca, A. H. Fox, the Winchester 21, and the Remington 32.

The Elsie was not a perfect gun; indeed, she was not as good as the best of her time. But she was graceful and elegant — and there were more than a few who loved her.◆

THE HENRY
Grandaddy of the Lever Actions

The foundation for a company that would eventually be known as Winchester, the Henry Rifle earned a unique place in the history and romance of the West.

RICK HACKER

Rarely in the annals of firearms history has one gun caught the romance and imagination of the public as universally as the repeating lever-action rifle. In both the real-life adventures of 19th century America as well as later in the re-written history of the silver screen and printed word, it symbolized the fast action, straight shooting American frontier spirit that came to eptiomize the "good guy" for the rest of the world.

By far the most famous of the lever-action rifles was the Winchester '73, the gun that purportedly "won the West." And Winchester's still-popular Model 94 currently holds the undis-

During the 1860s, the Henry's 16 shots made it the last word in firepower. This .40-44 replica Henry was produced by Navy Arms. Photograph by Art Carter

puted title of having the longest continuous production run of any mass-produced weapon. But both of these well-known lever guns, which flank opposite sides of America's frontier years, owe their existence to the granddaddy of the lever actions, the gun that started a trend — the Henry rifle.

The Henry was the first commercially successful repeating firearm to utilize fixed ammunition, and by the very virtue of this fact, it laid the foundation for a company that would eventually be known as Winchester.

Ironically, the story of the Henry rifle's success began in failure. In 1860, a fledgling east coast rifle manufacturing firm known as the Volcanic Repeating Arms Company became insolvent. Left holding a proverbial bag of gun parts and patents was one of Volcanic's major financial backers, a successful New England shirtmaker named Oliver F. Winchester.

Although not extremely knowledgeable regarding firearms design, Winchester was fascinated by guns. In an effort to salvage both his hobby and his rather sizable cash investment, he reformed Volcanic into a firm known as the New Haven Repeating Arms Company. To see if anything at all profitable could be made from the strange-looking little .36 caliber Volcanic with its odd-shaped lever and trigger guard combination, Winchester hired a talented designer

By merely flicking his wrist and pulling the trigger, a rifleman could unleash 16 rounds in 30 seconds. In one test, an Army marksman fired 120 aimed shots in 340 seconds, which included reloading!

named B. Tyler Henry to supervise his poor corporate orphan.

Henry did all he was asked and more. In an era that was still dominated by muzzleloaders, Henry focused his attention on the diminutive .36 caliber rimfire cartridge, which was prone to misfires and burst casings if it fired at all. Increasing case strength, powder capacity and striking force by enlarging the caliber to .44, Henry next redesigned the Volcanic to handle the new cartridge, which, in honor of its inventor, was called the .44 Henry Flat. The "flat" nomenclature was a subtle acknowledgement to the bullet's flat nose, which prevented it from making direct contact with the primer of the preceding cartridge which was designed to be carried in a tube underneath the rifle's barrel, a feature that the Volcanic had previously incorporated. However, the Volcanic's picket-shaped lead slug obviously did not have this flat-nosed advantage, thereby creating what was probably the first known cartridge gun version of the old cap and ball "multiple discharge."

The Volcanic was notable in that it possessed practically every adverse trait a gun of its type could have. The rifling often failed to grasp the bullet, resulting in keyholes at 25 yards. The cartridge itself was undersized and underpowered, approximating at best a striking force slightly greater than a gambler's vest pistol. Moreover, the Volcanic's extractor was subject to break and the gun often failed to chamber rounds within its cast iron frame. B. Tyler Henry changed all that by the outbreak of the "War Between the States," and his timing could not have been better.

Perfecting first his copper-based rimfire cartridge (which was stamped with a letter H on the base in honor of the inventor) and then the gun, the new Henry Repeating Rifle, for all its improvements, bore a surprisingly close resemblance to the old Volcanic. But there were differences. To handle the bigger .44 caliber cartridge, the gun was made larger and sported a brass frame, which was less costly and easier to manufacture than iron. The famous toggle-link action, which would be retained in the still-to-come Winchesters 66, 73 and 76, was both strengthened and simplified, resulting in a surprisingly fast action. To guard against potential misfires caused by "dead spots" in the rimmed priming compound encircling the case, the bolt on the Henry Rifle featured twin firing pins. With an eye towards the lucrative possibilities of government contracts on the eve of the great war, the gun was fitted with sling swivels on its left side.

Of course, the Henry was not without its faults. For one thing, it still retained the Volcanic's open parallel slit, which ran along the length to the bottom of the magazine tube. To load the Henry, the bullet plunger, which was compressed by a spring, was forced up towards the muzzle and the top portion of the magazine tube could then be rotated out to the side, thereby permitting the bullets to be dropped, base first, down the tube. The magazine tip would then be swung closed and the plunger allowed to rest atop the foremost cartridge, using its spring pressure to feed each successive round into the chamber with the opening and closing of the lever.

Using this system, which was kept on the Henry throughout its six years of manufacture, there were three ways in which the gun could be made inoperative, thereby taking it out of the repeater class and relegating it to the ranks of the single-shot: (1) it was a simple (and often occurring) matter to inadvertently trap snow, twigs or mud in the exposed slit in the magazine tube, thereby jamming the multiple shot capabilities of the gun, (2) a fall from horseback or a similar hard blow could dent the magazine tube entirely; secondary risk in this type of situation also meant that the rimfire cases could be damaged or worse yet, ignited while still in the tube and facing another cartridge directly ahead, (3) during rapid firing of the Henry, the magazine plunger gradually moved down, towards the shooter's supporting right hand. Thus, when the plunger reached a spot approximately five rounds from the chamber, the gun would fail to feed, unless the hand was momentarily removed and the plunger permitted to pass. In all fairness, however, it should be added that the exposed cartridges in the Henry's magazine tube did permit a rifleman to take immediate count of how many rounds he had left in his gun, a feature that the later Winchester lever actions could never offer, by virtue of their "improved" designs.

There was another factor that would have seemed to mark the Henry for failure, and that was the rather underpowered loading of the cartridge it was designed to handle. In an era when big-bore muzzle-loading plains guns were thundering forth with bullets measuring over a half-inch in diameter and backed by 120-plus grain powder charges, the 216-grain lead, pug-nosed Henry Flat spun out of its 24-inch barrel at 1200 feet per second, all the speed it could muster from its 26-grain black powder charge. Hardly a grizzly-stopper!

In spite of these apparent drawbacks, the Henry Rifle had one overriding fact that was to make it one of the most sought-after long arms on the early American frontier: firepower! By merely flicking his wrist and pulling the trigger, a skilled rifleman could unleash sixteen rounds — the total capacity of the rifle — in less than

30 seconds. By comparison, it took an average time of 35 seconds to load and fire a single shot from a standard issue Springfield rifled musket.

Of course, such rapidity of fire was unheard of, but with the impending war looming ever closer, skeptics were almost forced to become believers. In one government test, an Army marksman fired 120 aimed shots in 340 seconds, which also included the time necessary for reloading. But even though this averaged out to 2.9 seconds per aimed round, the Army refused to initially order the Henry. Objections were: unfamiliarity with its "new" design, fear of premature detonation of the cartridges while still in the magazine (no doubt a hold-over prejudice from the Volcanic rifle days,) possible unavailability of parts and ammunition in remote areas, and the relatively high cost of the firearm itself. In a day when good quality muzzleloading rifles could be purchased for $16 apiece, the price for a new, in-the-box Henry was a hefty $47.

In spite of the government's disapproval of the Henry Repeater, Oliver Winchester was not discouraged. He was a marketing man and through a series of promotions and aggressive sales force, he began to make the virtues of the Henry known. With the firing upon Fort Sumter in April of 1861, the head of the New Haven Repeating Arms Company was pleased to note that his revolutionary rifle was being privately purchased by many Yankees who could financially afford the lever gun. He must have also felt a surge of elation when, in 1863, the government ordered the first 250 Henry rifles of what would become a total U.S. order of 1,731 guns out of the entire run of 13,500 Henry rifles that were eventually manufactured.

The Henry immediately began making a name for itself in the war and soon became known as "...that damned Yankee rifle that can be loaded on Sunday and fired all week." The repeater was beginning to create its own legend, but it was a legend steeped in facts.

Rare photograph of Union Pacific Engine No. 23 and track construction crew. Note railroadman with his Henry.

One of the most celebrated stories involving the Henry rifle during the "War Between the States" involved a Union captain named James M. Wilson who lived in a part of Kentucky that was strongly for the South. As might be expected, Capt. Wilson's bluecoat leanings did not do much for community relations. Threats against his life became a regular occurrence. But Wilson had one thing in his favor: he owned a Henry rifle. The recounting of this true tale from the New Haven Arms Company sales brochure of the time says it best:

"...Captain Wilson had fitted up a log crib across the road from his front door as a sort of arsenal, where he had his Henry rifle, Colt's revolver, etc. One day, while at home dining with his family, seven mounted guerrillas rode up, dismounted, and burst into his dining room and

commenced firing upon him with revolvers. The attack was so sudden that the first shot struck a glass of water his wife was raising to her lips, when Captain Wilson sprang to his feet, exclaiming, 'For God's sake, gentlemen, if you wish to murder me, do not do it at my own table in the presence of my family.'

"This caused a parley, resulting in their consent that he might go outdoors to be shot. The moment he reached his front door, he sprang from his cover, and his assailants commenced firing at him. Several shots passed through his hat, and more through his clothing, but none took effect upon his person. He thus reached his cover and seized his Henry rifle, turned upon his foes, and in five shots killed five of them; the other two sprang for their horses. As the sixth man threw his hand over the pommel of his saddle, the sixth shot took off four of his fingers; notwithstanding this he got his saddle, but the seventh shot killed him; then starting out, Captain Wilson killed the seventh man with the eighth shot."

Other documented tales of the Henry were no less heroic, or less in favor of the Henry; in March of 1863, fifteen men from Col. Netter's Kentucky Volunteers, all of them armed with newly-purchased Henrys, successfully beat back a charge of 240 musket-armed Rebels. In another incident, in a moment of what must surely have been combined inspiration, every member of the First D.C. Cavalry turned in his issue .58 caliber Springfield and went out and bought a Henry rifle. Later their commander would state that the Henry "...could easily put two balls out of three inside a ring two feet in diameter from six to seven hundred yards."

While such accuracy of the Henry was occasionally possible, it was clearly not designed as a long-range rifle. Besides, the striking force still retained by its underpowered cartridge at those extreme ranges leaves its effectiveness as a long-range sniper's rifle open to speculation. Its chief advantage was psychological and years later, even after the advent of the Winchester 73s and 86s, many Indian tribes, most notably those in the Southwest, considered the Henry a highly desirable war trophy.

The ending of the war in 1865 did not end the Henry's career. Rather, it began a whole new chapter. Like so many ex-war guns, the Henry went west and blazed entire new legends, thus setting the stage for the Winchester guns.

One of the earliest post-war stories involved two Union troopers who were mustered out, but managed to keep their cherished Henry rifles. It was lucky for them that they did, for while prospecting in the northwestern area of the Rockies, they were surrounded by a war party of Blackfeet. As far as the Indians were concerned, this was still the age of the single-shot muzzleloader and consequently, their war strategy involved having a number of braves show themselves as targets. When each white man had fired his gun, the rest of the party would attack before the victims could reload. In this case, maneuvering into extremely close range, two Indians, on schedule, leapt straight up out of the grass and both troopers fired. The balance of the braves attacked but the men kept on firing. Confused and with other members of their tribe falling all around them (for at that range, both white men later recounted, it was hard to miss,) the Blackfeet turned and retreated before all 32 shots from both Henrys could be consumed. But in what has become a classic frontier tale of the Henry's firepower supremacy, the white men kept on firing. They then reloaded and filled every fallen warrior with a multitude of bullet holes, just so there would be no mistake as to what the terrible "Spirit Gun" (as the Indians thereafter called it) was capable of. The men were never again bothered by Blackfeet as long as they stayed in the Rocky Mountains, and they would take pleasure, sitting in front of their cabin, watching the Indians circle great distances out of their way in order to avoid setting foot on the troopers' property.

During this time of Western expansion, the Henry rifle was purchased by many railroads, most notably the Union Pacific, in the epic struggle to link the East and West coasts by rail. The UP surveyors, often 300 to 400 miles out in front of the main track-laying party, were extremely vulnerable to Indian attack. In an effort to provide their crews with the most up-to-date firepower available at the time, UP armed a number of these men with Henrys, which were stamped UPRR on the left side of the receiver. They are among the rarest Henrys.

With the introduction of the Improved Henry Rifle (a name that was later changed to the Winchester Model 1866), the Henry was discontinued, although a number of them still found their way into the hands and saddle scabbards of the early explorers and pioneers. But the cartridge had already been superseded by the .44-40, which was busy blazing its own path to immortality. The Henry might have been all but forgotten, were it not for the collectors and firearms historians of today.

Currently, any original Henry rifle commands a premium among lever-action aficionados. Depending upon condition, geographical location, and general attitude of the purchaser and seller, a Henry rifle can bring anywhere from $2,500 to $15,000, with the latter price being reserved only for those guns in pristine condition or with a unique historical association. However, a few years ago, I did see a foreign copy of a Henry sell for $9,000! Rarest of the Henrys are the very early iron-framed models, the chief difference being the sharper profile of the buttstock in the later gun. Serial numbers, which can be found on the top flat of the barrel, just behind the rear sight, give an approximate indications of year of manufacture (although some overlapping does exist), per the chart below.

Interestingly, both Navy Arms (689 Bergen

Approximate Year of Henry Rifle Production (missing 1860-61)		
YEAR	No. of Guns Made That Year	Serial No. at Year End
1862	1500	1500
1863	2500	4000
1864	4000	8000
1865	3000	11,000
1866	2500	13,500

The scarce military models are stamped with the inspector's initials "C.G.C." on the barrel near the breech and on the stock.

Blvd., Ridgefield, NJ 07657) and Allen Arms (1107 Pen Rd., Santa Fe, NM 87501) have recently introduced replicas of the Henry. Navy Arms limited their replica to 500 units, which initially sold for $500 each and now have reappeared on the collector's market for anywhere from $850 to $1,500, but current value is not obtainable due to the low number of guns in existence. One of the rarest of the replica Henrys is the UPRR Commemorative, of which only 100 were made. The guns all sold out within two days for the retail price of $750, and a recent price offered for the prototype (which was subsequently turned down) approached the value of original guns. Although all versions of the Navy Arms Henry are now sold out, a new limited edition military commemorative will be offered by Navy later in 1982.

Allen Arms is importing an excellent copy of the Henry rifle, which, like Navy's will sell for $500 per gun. However, unlike Navy's, these replicas will not be limited editions and will become a regular part of the Allen Arms line.

All of the replica Henrys can be found chambered for the "proper" .44 Rimfire (a cartridge which has been obsolete for a number of years but is now reportedly being brought back by Navy Arms) or the more readily obtainable and shootable (but historically inaccurate) .44-40. It should be pointed out that both of these replica-producing firms also offer a .44 Henry Carbine, but such a gun never existed. Nonetheless, with the originals gaining in value and the replicas becoming an enticing product for both shooter and collector, it seems that the legend of the Henry rifle is not over yet.♦

FORGOTTEN
Classics

Remington's doubles pioneered the great age of American shotguns. Today, they are relatively obscure among shooters and collectors, overshadowed by their showier counterparts.

MICHAEL McINTOSH

Immaculate engraving graces the receiver on this 1894 Remington EE. Photograph by Art Carter

Eliphalet Remington himself built the first shotguns to bear the Remington name. Almost from the beginning of a career that started in 1816 with a flintlock rifle built in his father's smithy in central New York State, the man who had dreamed of being a poet turned out single-barrel guns smooth-bored for shot. The earliest, naturally, were flintlocks, followed in due course by caplock designs. By the 1840s, when the business had grown into a fully-staffed factory along the Erie Canal, there were side-by-side combination guns, one barrel rifled, the other smooth, available custom-built for the sportsman who wanted to be ready for anything.

But it was not until well after the War Between the States that shotguns truly found a place among Remington's annual output of rifles, revolvers, pocket pistols, and carbines. Still, Remington was among the first important American makers to give the shotgun market a serious look, and when the Remington breechloader premiered in 1874, the great age of the American shotgun had barely begun.

In an unevenly populated country less than 100 years old and still a generation or more from even establishing its geographic boundaries, the rifle naturally was king, both for survival and for sport. What slender share of the gun market that the shotgun could sustain was, until about 1880, largely supplied by guns imported from Europe, where the breechloading double was already reaching the peak of its evolution. Of the American guns that would achieve lasting fame, only the Parker existed by 1874.

So the 1874 Remington was an auspicious item, not only for its early appearance but more important, because it was an excellent gun that could be built and sold for a price substantially lower than that of its European counterparts. With few exceptions, that principle characterized products of the American shotgun trade for nearly 100 years after.

The gun was built according to patents issued to Andrew E. Whitmore (or Whittmore, as it's sometimes spelled) in August 1871 and April 1872. In most respects, it was a typical shotgun of its day — a break-open breechloader with tall hammers and sideplate locks — though there was one unusual feature. The fastening system operated by a top lever, but rather than pivoting from side to side, Whitmore's latch pivoted up and down. The shooter pushed the top lever up, presumably with his right thumb, to withdraw the bolt, a typical sliding underlug, from its notch in the barrel lump. The idea worked and it remained in production until 1889.

The Remington-Whitmore was made as a shotgun in both 10 and 12 gauge, as a rifle/shotgun combination, and even as a double rifle. The standard shotgun, which sold for $45 in 1877, had steel barrels and a plain walnut stock. The middle grade, with twist barrels and figured wood, cost $60; for $75 you could have Damascus barrels, fancy walnut, and some engraving on the frame and lockplates. In a revised, 1878 version there were two additional grades, one at $85 that featured English walnut and a $100 Superior Grade with fine Damascus barrels, English-walnut stock, and elaborate engraving. For a few dollars extra, you could order a pistol-grip stock or, after 1878, rebounding locks.

The Whitmore gun apparently was a grand success. Actual production figures are long gone, and considering that the shotgun market in America was not brisk at the time, a widely-accepted estimate of 50,000 guns produced between 1874 and 1878 is probably too high. Still, there is no question that sportsmen found the Remington an excellent value. In 1882, the Whitmore design was revised again and two more grades added, bringing the total to seven. Prices ranged from $30 to $100. Further revisions were made in 1885 and 1887, when 16-gauge guns appeared for the first time. But the Remington hammer gun reached its ultimate form with the model of 1889.

The catalogues called it the Remington New Model, just as they had each version since 1878. This time, though, it really was new, a substantially different gun from the one Andrew Whitmore had designed nearly 20 years before. For one thing, the lifter-type latch was gone, replaced by a conventional top lever. The bolting system, too, was different, with a doll's-head rib extension to supplement the sliding underbolt. The hammers, which are of a somewhat different shape in each of the revised models, are by now low-profiled with nearly circular necks and spurs much shorter than those of the original Whitmore guns.

There were seven grades, ranging in price from $30 for a No. 1 to $125 for the elegantly finished No. 7. The 10 gauges were available with barrels 30 or 32 inches long, the 12s and 16s with tubes of 28, 30, or 32 inches. All Model 1889 guns were proofed for smokeless powder, which suggests that Remington was among the earliest American gun-makers to realize that black-powder days were coming to an end. Still, typical of the time, the highest grades had Damascus barrels, and most Model 1889s, regardless of what sort of barrels they had, were bored for 2⅝-inch shells.

The Model 1889 was unquestionably the best of Remington's hammer guns and, in fact, was one of the best hammer guns in America. It was sturdily built from good-quality materials and, in the lower grades especially, was a far better gun for its price than the Belgian imports and cheaply-made domestic guns that began to flood the shotgun market at the end of the 19th century. In its last years, it was a mainstay of the mail-order trade; the 1908 Sears catalogue offered the No. 3 Grade, which had two-blade Damascus barrels and a case-hardened frame, at $23.25 in either 10 or 12 gauge. The same gun listed at $35 in the factory catalogue.

The 1889 remained in reproduction until 1908, only two years before Remington stopped building side-by-side double guns of any type. Total production may have been as great as 100,000 guns — not a bad showing for an item that in its heyday competed directly with similar offerings from Parker, L. C. Smith, Ithaca, and a host of lesser lights.

But the gunning world was headed in a different direction and by the last decade of the century, the hammer gun was well on its way toward becoming a fossil.

Interest in hammerless actions, which flourished in England in the 1870s, gradually reached across the Atlantic. At the time, Dan Lefever seems to have been the only important American marker to take the idea seriously; his 1878 hammerless breechloader was one of the first such guns made in this country. But the change came slowly and when Lefever's masterpiece, the Automatic Hammerless, appeared in 1885, it was still the only high-quality, American-built hammerless gun. The first L. C. Smith hammerless went into production a year later, followed in turn by similar designs from Parker in 1889, Baker in 1890, and Ithaca in 1892.

Patented in 1871 and 1872, the Remington-Whitmore was made as a shotgun in both 10 and 12 gauge as a rifle/shotgun combination, and even as a double rifle.

Remington, once in the forefront of American shotgun-making, now found itself lagging behind. The 1880s were difficult years. After Eliphalet Remington's death in 1861, ownership and management of the firm had passed to his three sons — Philo, Samuel, and Eliphalet III. Though Philo acted as general manager, Samuel, as sales manager, seems to have been the key force in the company's success. Samuel died in 1882, and the financial fortunes of Remington & Sons began to decline. The two remaining brothers declared the firm bankrupt in 1886. Still, both the quality of Remington's products and the demand for them remained high.

Marcellus Hartley, an astute businessman with a background in manufacturing and retailing of sporting goods, bought the company from the receivers in 1888. Hartley had been one of the founders of Union Metallic Cartridge Company, and he immediately combined Remington's sales program with UMC's, keeping both the guns and manufacturing techniques as they had been.

One of the best hammer guns in America, Remington's Model 1889 was advertised in the 1908 Sears catalogue for $23.25, complete with Damascus barrels and case-hardened frame.

Remington Arms, as Hartley renamed it, was off and running once again.

Marcellus Hartley in the 1890s was free from the internal turmoil that had wracked the Remington brothers in the '80s. He undoubtedly realized that Remington could claim a significant share of the burgeoning shotgun market. The firm's reputation was already in place, and the Model 1889 hammer gun was a popular item. For the double gun, though, the hammerless action clearly was the future, and Hartley ordered design work to commence.

The gun appeared on the market as the Model 1894, a sturdy boxlock of the Anson & Deeley type in which leverage from the barrels, pivoting on the hinge pin, lifts the internal hammers into their sear notches. The bolting system, operated by a top latch, comprises twin underlugs assisted by a third bolt that bears against a rib extension.

All told, the Remington hammerless is as reliable and strong as any American gun.

Like the hammer guns, the standard Model 1894 came in seven grades, three gauges, and with a vast variety of options. The grades began with K as the lowest and progressed upward through A, B, CE, DE, EE, and the magnificent Remington Special, which in 1902 fetched the hefty sum of $750 — about the same price as the L. C. Smith A3. By comparison, the highest-grade Parker of the time, the AA Pigeon Gun, sold for $400. In both the hammer and hammerless guns, the highest grades were built on special order only. These included grades 4, 5, 6, and 7 in the hammer models, and grades CE, DE, EE, and Special in the hammer guns. Any high-grade Remington double is an extremely rare item today.

The Model 1894 also was made as a trap gun, and that version had its own grading system, comprising grades F, C, and D. These essentially correspond to the B, C, and D grades in the standard line.

Automatic ejectors were a $5 option. All grades could be had with either solid-steel or, for $10 more, twist barrels. Ten-gauge guns came with 30- or 32-inch barrels; 12-gauge weights ranged from 6¾ to 8¾ pounds, 16s from 6½ to 7½ pounds. There is no evidence that Remington ever built doubles in gauges smaller than 16.

Beyond the basics, a customer could choose among a candy-store array of options, including multi-blade Damascus barrels, flat or hollow ribs, and fine English-walnut stocks in straight-hand, half-hand, or pistol-grip style. Buttplates could be hard rubber, horn, or skeleton steel. An automatic safety was standard, but it could easily be converted to manual mode by taking off the buttstock and removing the safety plunger. Extra sets of barrels, for any Remington double, cost half the price of the gun they were made for.

Double triggers also were standard fare, but Remington eventually owned patents on at least two single-trigger designs — issued in 1902 and 1906 — so it's possible that some guns in the last years of production were built with them.

Remington grade markings often indicate more than just the amount of decoration. Most makers added an "E" to the grade stamp of an ejector gun, but Remington went a step further and included the type of barrels as well. For instance, a K Grade with Damascus barrels usually was stamped KD. The same gun with ejectors and best fluid-steel barrels (which Remington called Ordnance Steel) might be stamped KEO. A gun barrelled with a cheaper grade of fluid steel — called simply Remington Steel, and also guaranteed for nitro powders — would be stamped KER.

At the turn of the century, Remington made some mechanical revisions in its hammerless design and called the results the Model 1900, a gun virtually identical in appearance to the Model 1894. The Model 1900 was meant to be an economy gun; it was built only in 12 and 16 gauges, and only in plain grades. Both models remained in production simultaneously.

But not for long. Marcellus Hartley died of a heart attack in 1902 and was succeeded by his grandson, Marcellus Hartley Dodge, a young man who seems to have seen the future with remarkable clarity. He predicted, among other things, that repeating shotguns would one day dominate the market, a fact that Remington's arch rival, Winchester Repeating Arms, was slower to accept. Dodge moved quickly. By 1905, Remington was manufacturing John and Matthew Browning's square-backed autoloader as the Model 11 and in 1907, brought out a pump gun called the Model 10, designed by John Pedersen. Not only did such guns offer the firepower that would so captivate the American gunner, but they also could be built in great numbers without the expensive hand-work required for a double. Choosing to cast its lot entirely with mechanical guns, Remington discontinued the Model 1889 hammer gun in 1908 and its hammerless models in 1910.

Remington's decision was economically shrewd. It anticipated the overwhelming trend of the markets and allowed the factory to concentrate on guns that could be built almost entirely by highly-mechanized, mass-production techniques. Remington dominates the American arms industry today for precisely the same reasons.

But in terms of the artifacts of history, there was a price, and that decision years ago consigned some good guns to a premature obsolescence. Among shooters and collectors alike, the Remington doubles remain unhappily obscure. In the broad genre of American doubles, they are far overshadowed by Fox, Smith, Parker, Ithaca, and others. That they aren't better known and better appreciated seems to me an unfortunate quirk of events. Certainly, it has little to do with intrinsic merit, because the Remingtons can make a good showing for themselves in any company. That so few were built as high grades didn't help, either; among American guns especially, reputations often rest more on cosmetic appeal than on the quality that lies beneath the surface. Unfortunately, the Remington doubles disappeared from the public eye before the great years of the American double, the years when enduring reputations would be made. But history, like life, is seldom fair.♦

Mystique of the MODEL 70

The pre-1964 Model 70 Winchester is the most coveted rifle in the world. A paragon of design and finish, it richly deserves the title, "Rifleman's Rifle."

DAVID E. PETZAL

Its name, in the shorthand of the sportsman, sounds like something you might hear on a football field — pre-64 70. Its manufacturer, Winchester, called it "The Rifleman's Rifle." During the golden years of big-game hunting, it was *the* big game rifle.

However, it is a rifle of contradictions. The Model 70 was not an original design. By today's standards, it is both heavy and inaccurate. Yet 18 years after the last of the original Model 70s were produced, it remains a paragon among rifles. It can be found in Alaska and in Africa; in Montana and Michigan and Mississippi. The original Model 70 action has only one serious rival in the fine custom rifle field, the Mauser. The Model 70 is also becoming one of the real prizes of this decade for collectors.

The ancestor of the Model 70 was the Winchester Model 54. Introduced in 1925, it was patterned closely on the 1903 Springfield and was moderately successful. The 54 was smooth and positive in operation, yet it suffered from several flaws: the sear doubled as the bolt stop, which detracted from the trigger's quality; the high profile of the bolt and safety made scope mounting difficult; and the trigger guard was a decidedly unattractive steel strap.

Winchester was not satisfied with the 54, and commissioned two of its designers, Leroy Crockett and Albert Laudensack, to improve the gun. Under the supervision of chief designer Edwin Pugsley, they created a work of genius. Christened the Model 70, the new design was completed in 1934 (although the first rifles were not actually shipped until August 1936).

The Model 54's heavy non-rotary extractor, receiver-mounted ejector and partial bolt face-rim were retained in the new design, as was the coned breech. This feature, actually a holdover from the Springfield, is one point cited by detractors of the Model 70. The rear of the Model 70 barrel is cone-shaped when viewed in cross-section, and while it greatly facilitates the feeding of cartridges, it weakens the

Photograph by Art Carter

breeching of the gun. In combination, these devices made for an incomparably smooth and certain mechanism.

The Model 54 bolt handle was replaced by one which was not only lower, but is considered by many to be the handsomest ever designed. Gone was the steel-strap trigger guard. In its place was a solid steel trigger guard and a hinged steel floorplate, which improved appearance and made unloading the magazine far more convenient. A low, pivoting safety was utilized as well.

Perhaps the crowning achievement of Crockett and Laudensack is the Model 70 trigger, which has never been improved upon and, according to many shooters, never equaled. It is characterized by extreme simplicity and ruggedness. When properly set up by a gunsmith, the trigger provides a clean, crisp, almost motionless break, and will probably never have to be touched again in the shooter's lifetime. It is one of the very few triggers that will stand the pounding of competitive big-bore rifle shooting, which makes the pre-64 Model 70 highly favored by that clan.

The Model 70 was, however, a complex action to build, and in this complexity lay the seeds of its demise. The receivers were machined from massive bars of chrome-moly steel, in a process that took approximately 75 cuts. Other, more modern actions, are machined from round bar stock and require only a fraction of the work.

However, in 1936, good machinists and gunsmiths were happy to be working, and so the Model 70 had a bright future. At this point Winchester could have taken its splendid new action and incorporated it into a rifle that was poorly put together, had mediocre lines, or shot badly; but such was not the case. The Model 70 was a graceful gun, handsome in every way.

By the standards of the time it shot superlatively and, perhaps most important, it was a gun that was built with very obvious care. Gary Herman of Safari Outfitters in Ridgefield, Connecticut, sums it up best: "The pre-64 Model 70 was a mass-produced custom gun."

Anything this good could not help but succeed, even in the Depression. The shooting public took the Model 70 to its heart, and Winchester responded by offering the rifle in just about any shape and form imaginable.

Model 70 production can be broken down into three "periods."

The first period extended from 1937 until 1942, when the demands of World War II forced Winchester to manufacture only military arms. About 50,000 "pre-war" Model 70s were made during these years, mostly in .30/06; they are regarded as the finest of the lot.

Sometimes called the "Transition," the second period extends from the end of 1945 until 1951. Approximately 160,000 second period Model 70s were produced, again with the .30/06 as the most common caliber. They can be recognized by the "Transition" safety and straight receiver tang. The quality is high but is a distinct cut below that of the earlier rifles, as Winchester had begun cutting back on handfinishing.

The third period runs from 1952 until 1964, with the introduction of the new Model 70. It was during this era that the variety of models was increased. The Featherweight was introduced for hunters who wanted a lighter hunting rifle. Cartridge families (.243/.308/ .358, for example) were used as the basis for design. This was also the period when magnum cartridges (.458, .338 and .264 Winchester Magnums) were introduced to an eager market. To add to their sales appeal, the rifles chambered for these cartridges were given recognizable names: Westerner (.264), Alaskan (.338) and African (.458). Production total for the third period was 360,000 rifles, but quality was in a steady decline. Rifles of this era, with the exception of the Super Grade (which was discontinued in 1960), are still perfectly sound firearms, but the quality of the wood is only average, and the inletting and checkering are fair to poor. I know more than one gunsmith who had a thriving business during the 1950s-60s correcting the factory flaws in Model 70s.

In 1963 the ax fell after about 583,200 Model 70s had been produced, and in 1964 a new model 70 was introduced. It was as poor a rifle as its predecessor had been great, and it was thoroughly despised by the shooting fraternity. Since then the gun has been improved to a state of respectability. Thanks to good thinking by the management of Winchester and U.S. Repeating Arms (who now makes the Model 70), current Model 70s are as well made and tastefully designed as any current factory made, high-powered rifle.

The Model 70 was available in four basic types: Standard, Competition, Deluxe or Super Grade, and Custom (built to special order by the Winchester Custom Shop). Each model type contained a number of styles. Standard included: Standard Rifle, Standard Carbine,

Jack O'Connor (right) and guide with Idaho bull elk taken with a Winchester Model 70, his favorite rifle.

Standard Featherweight rifle, Varmint Rifle, Alaskan and Westerner. Competition accounted for the Sniper's Match Rifle, the National Match Rifle, the Target Rifle and the Bull Gun. Deluxe or Super Grade took in the Super Grade Rifle, Super Grade Carbine, Super Grade Featherweight Rifle and African model. The Custom category included special-order guns. Confusing, yes. But it is the very diversity of the Model 70's numerous shapes and sizes that makes it so attractive to collectors.

Winchester could have taken its new action and incorporated it into a rifle that was poorly put together, had mediocre lines, or shot badly; but such was not the case. The Model 70 was a graceful gun, handsome in every way.

As if this were not enough, the Model 70 was chambered for a wide variety of calibers, and it is the rarity of certain calibers that makes some 70s extremely valuable. There are 20 regular chamberings, and another 19 that were either made on special order, employed in experimental guns, or rumored to exist. The standard chamberings, from most common to rarest, are as follows: .30/.06, .270, .308, .243, .300 H&H, .375 H&H, .220 Swift, .257 Roberts, .22 Hornet, .264, .458, .338, .250/3000, 7x57, .300 Winchester Magnum, .358, .35 Remington, .300 Savage, 9x57 Mauser and 7.65 Argentine.

For the collector, acquiring a good Model 70 at a sane price will require some work. Remember that, with the exception of the ultra-fancy custom shop guns, the Model 70 was a working firearm and produced in large numbers. Unlike fine shotguns which were usually treated tenderly, the Model 70 was used hard, often abused or altered, and re-sold time and again.

On top of that, there has been a lot of faking. It is far easier to alter a Model 70 than a Purdey. Tampering with a fine old shotgun to increase its value requires the services of a highly skilled gunsmith but cobbling a Model 70 can often be done with remarkable success by an amateur.

So before you go thundering off in search of a bargain-priced bonanza, I urge you to buy the revised edition of *the Model 70 Winchester 1937-1964*, by Dean H. Whitaker. Whitaker has assembled a truly awesome amount of information on the Model 70, and organized it into comprehensible form. This book is aimed at the collector rather than the historian, and a considerable number of pages are devoted to the more likely fakes that you may encounter. *The Model 70 Winchester 1937-1964* is available from Saromi Press, Box 261, Los Alamos, N.M. 87544, and costs $24.95. I cannot overemphasize the value of this publication if you plan to be a serious collector.

Hazards aside, the Model 70 is an outstanding collectible. It has brought good prices for a number of years, and continues to sell well despite the current recession. Part of the reason for this is that even the rarest 70s have yet to come anywhere near the princely figures commanded by other valuable firearms, so people can actually buy them.

For example, according to Griffin & Howe, a Super Grade .22 Hornet (Transition period) in excellent condition will bring about $1,500. The most expensive Model 70 ever sold by Safari Outfitters is a Super Grade transition model in 7x57. It went for $3,500, and that was two years ago. This is not what you'd call grocery money, but when you contrast it with the $15,000 to $60,000 that dealers are now asking for a really desirable shotguns, it is quite a bargain.

What do you look for? Well, almost any pre-64 Model 70 is worth something, if only for its action, which is beloved by custom gunsmiths. So if you find a Model 70 with a battered stock and worn-out barrel, don't ignore it. If the action is sound, it is still worth anywhere from $250 to $300.

The most desirable rifles are the ones in mint or near-mint condition. This means perfect, like new, with not a mark on them. Premium prices are paid for a Model 70 still in the original factory box with tags and labels. These latter specimens are the ones that collectors crave. Genuine factory-engraved Model 70s are also high-ticket items, but extremely rare.

If you find a specimen that has a few minor blemishes, resist the temptation to fix them yourself. In some cases, condition is not the overriding factor. An unused, in-the-box standard grade .30/06 circa 1962 is worth far less than a battered, pre-war Super Grade .35 Remington. Caliber is probably *the* most important part of the equation for determining the value of a Model 70, although period, grade, special features, and condition must also be figured in. It is, however, that unknown factor in the collector's game (*who wants it and how bad*) that remains the most important variable.

I'd like to be able to tell you to watch the advertisements in *Shotgun News* and hit the gun shows to pick up a real value, but I don't think this is good advice. Most people who advertise in the former or attend the latter are going to have a pretty good idea of what is worth how much, and you're not going to get many bargains. My advice is to drop in on the small gun stores, watch the bulletin board at your local range, and covetously eye what your fellow shooters pull out of their gun cases.

The small gun stores can be a gold mine. In a recession, people sell guns to raise cash. Often the proprietors of these stores will know vaguely that a Model 70 is worth money, but they usually have no idea how much. In one such instance, a local hunter decided that the .338 Alaskan, which first appeared in 1958, was just the thing for whitetail deer. He discovered, however, that pulling the trigger was no fun at all. After firing maybe 20 shots, he put the gun in the rack where it sat for 24 years until he decided to retire and go to Florida. The .338 went to the dealer with a $400 price tag. It is worth about $900, but our friend had brought it for $125 when it came on the market, and just wanted to get whatever he could for it and go fishing.

Perhaps that is the real appeal of collecting the Model 70 — you can expect to unearth a bargain, whereas your chances of finding a Parker A-1 Special or a Boss in someone's attic are about nil. The Model 70 was the custom rifle of the common man, and there are a lot of them still out there, waiting to be given the appreciation they deserve. "The Rifleman's Rifle" they called it; and not for nothing.♦

For some excellent technical background on the Model 70, I recommend:

The Bolt Action, by Stuart Otteson, Winchester Press, New York, 1976.

Bolt Action Rifles, by Frank de Haas. Digest Books Incorporated, Northfield, Ill., 1971.

The History of Winchester Firearms, by Duncan Barnes. Winchester Press, Piscataway, NJ, 1980.

The Rifleman's Rifle, by Roger C. Rule. Alliance Book, Inc., Northridge, CA 10982.

SECOND TO NONE

Though not "London Best," a host of other English gun makers, including W&C Scott, John Dickson & Sons and John Wilkes, did much to bring the shotgun to perfection.

MICHAEL McINTOSH

The sporting shotgun, as we know it, is largely a British invention. Though the origins of the smooth-bored gun reach back to the ancient dynasties of China and though the major phases of its evolution took place in Europe, the British ultimately defined both the shotgun and the sport for which it is used. In the process, they transformed a clumsy, unreliable weapon into a work of art.

While German, Italian and French armorers were evolving the matchlock into the wheellock into the flint gun, the English hunter relied upon a variety of nets, snares and springes. Guns, heavy and slow to fire, were of little use in taking game. Sport was the prerogative of the nobility, and was pursued with the hound, the hawk and the longbow. But with the Restoration came a new influence. When Charles II returned to England in 1660 from exile in France, he brought a slender flintlock fowling piece and a strange new sport — shooting birds on the wing.

Under Charles's patronage, wing-shooting caught the fancy of the English gentry. It was fine sport, a challenge of skill. English gun makers quickly set about copying Charles's French flintlock, and their improvements in design led to the first truly specialized bird guns — lighter, slimmer, stocked to fit the shoulder and the cheek. By the end of the 18th century, both the sport and the shotgun had reached a peak of refinement in England matched nowhere in the world.

The English countryside, as much as anything, affected the development of the sporting shotgun. It was a man-made landscape of small fields and tiny woodlands, ideally suited for small game. Its overall size and dense human population made it equally unsuited

Present-day "round action" John Dickson & Son shotgun with a portrait of founder John Dickson and two trade labels.

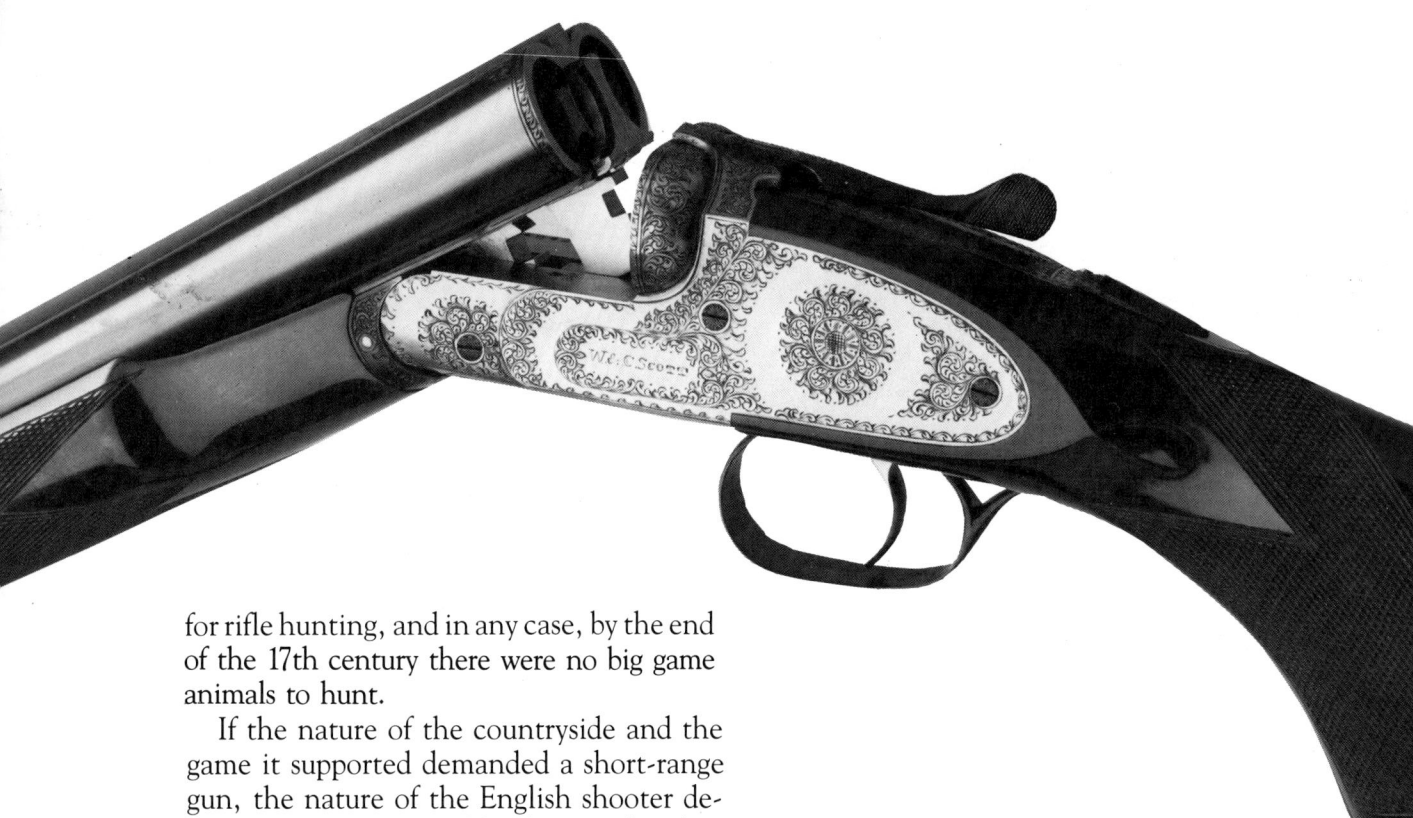

for rifle hunting, and in any case, by the end of the 17th century there were no big game animals to hunt.

If the nature of the countryside and the game it supported demanded a short-range gun, the nature of the English shooter demanded a fine one. Unlike the British military, so hidebound by conservatism and tradition that British cavalry troops launched saber charges against Boers armed with Mauser rifles as late as 1902 (with predictable results), the English sportsmen of the 18th and 19th centuries were by and large eager for new developments in sporting arms and were willing to pay handsomely for them. Shooting was a gentleman's pastime, and because the English gentleman's goods were, and often still are, made to order, gun makers took their place among the other London bespokemen — hatters, tailors, bootmakers and others.

Heir to such social and economic support, it is small wonder that the greatest English gunsmiths have always been shotgun makers. From the Manton brothers of the 18th century to the present day, the standards by which shotguns worldwide are judged bear the great English names: Lancaster, Purdey, Boss, Greener, Richards, Holland, Woodward, Churchill. The "London Best" gun is by repute the world's best.

In this century, the majority of English guns have evolved along essentially similar lines. Unlike the American gun-making industry, which trades largely on differences in design, the English have grown less inclined to change a good thing simply for the sake of change. There is less mechanical difference, for example, between a modern Purdey and a Holland & Holland than between a Parker and L. C. Smith. As more and more of the patents covering basic designs expired, English guns became more and more mechanically similar. Design principles which proved most reliable and most aesthetically pleasing have been incorpo-

The Texan by W&C Scott. W&C Scott has recently become part of the Holland Group. Photo courtesy of W&C Scott (Gunmakers) Ltd.

rated by most English makers. Today, what distinguishes one from another, particularly among London Best guns, often are refinements rather than significant differences.

But it wasn't always so. The last half of the 19th century, when the muzzle-loader gave way to the breech-loading gun, saw a revolution in arms making, and in those years a host of English gun makers helped bring the sporting shotgun to perfection. Some of the companies they founded and many of the guns they built have survived well into the 20th century; if their names are not now synonymous with London's Best, the quality of their work hardly justifies calling them second-best.

The great revolution began in 1851 at the Great Exhibition in London, when Casimir Lefaucheaux of Paris displayed his breech-loading pinfire shotgun. Lefaucheaux had been building hinge-action guns for nearly 20 years, but neither they nor the advanced pinfire cartridge which another Frenchman, Houllier, had developed in 1846 had elicited much attention from the English makers. But when Joseph Lang, gun maker at 7 Haymarket, London, saw it at the Exhibition, the Lefaucheaux gun struck him as having vast potential. When the shooting season began the following year, Lang offered his own guns built in the Lefaucheaux image. Both sportsmen and other English gun makers took notice.

The great Charles Lancaster of London took the breech-loader a step further in 1852 with improvements in both the bolting mechanism and the cartridge design. Lancaster's action was operated by an under-lever which, when swung to one side, cammed the barrels forward slightly before they pivoted down on the hinge pin. James Dougall, who had shops in both London and Glasgow, used a similar principle.

Lancaster's greatest contribution, however, was his cartridge, a brass-headed affair which functioned much like a modern rimfire. It not only helped solve the problem of gases leaking back into the breech but also provided better means of extraction. Refinements of the idea opened the way for modern center-fire ammunition.

By 1860, creative energy was flowing through the English gun trade like an electric current. A great many of the ideas led to blind alleys, but some were milestones. The second Thomas Horsley of York entered a patent in 1862 for a bolting system in which a single spring-driven underbolt fastens the barrels to the breech. The earliest design featured an under-lever; this was changed to a top lever in a patent issued in 1863. A third Thomas Horsley patent, issued in 1867, covered a mechanism by which the firing pins rebounded back into the frame after the cartridge was fired, thereby eliminating the need to pull the hammers to half-cock before the gun could be opened. The idea was perfected two years later by a lockmaker named Stanton of Wolverhampton, Staffordshire; in Stanton's system, the hammers themselves rebounded to half-cock after striking the pins.

Even as the classic exposed-hammer shotgun approached the zenith of its perfection, its successor appeared. The search for the best means of bolting an action all but ended when Purdey's patent on the double underlug expired and virtually every maker in England adopted the system. With that problem out of the way, attention focused on what to do about getting the hammers off the outside of a gun and into its action body.

Actually, concealing the hammers inside the frame was a relatively simple matter. How best to cock them was something else again. The most successful of the early designs featured a lever under the frame. Theophilus Murcott's gun was the first of this type to achieve real commercial success, and according to W. W. Greener, Murcott was a primary influence in the earliest popularity of hammerless guns in general. Perhaps so, but he was far from alone in the limelight.

George Gibbs and Thomas Pitt, both of Bristol, entered a joint patent in January 1873 for "improvements to Breech Loading firearms." Though their ideas included self-cocking guns of both exposed-hammer and hammerless types, some of the earliest Gibbs & Pitt guns featured a system in which the under-lever simultaneously retracted the action bolts and cocked the locks. In these guns, the lockwork is fastened to the trigger plate. A later design put the locks on more conventional-looking sideplates.

The most successful of the Gibbs & Pitt inventions was an adaptation of their original design to a top-lever system. This gun, which to all outward appearance is a typical modern sidelock, actually incorporates two distinct means of cocking the locks. As in the earlier system, the fastening bolt is linked to the lockwork, and if the top lever is moved fully to the right, the internal hammers are lifted to full cock. But if the lever is moved only far enough to retract the fastener, a cam mounted on the barrel lump pushes the bolt fully to the rear as the barrels drop down, permitting cocking leverage to be applied by either of the shooter's hands or by both hands together. A number of these Gibbs & Pitt guns were built by Bentley's and marketed by George Daw of Threadneedle Street, London, who earlier had also been a successful gun maker.

In 1875, William Anson and John Deeley, gunsmiths employed by the firm of Westley Richards, patented their design for a hammerless gun that was cocked by the leverage of its barrels, an idea that ultimately would become the standard system for the break-action gun worldwide. But the under-lever was far from dead. In the same year, Edwin Hughes entered a patent for an under-lever system that worked equally well with exposed hammers and those enclosed in a shotgun's frame. The firm of Joseph Lang & Son, founded by the man who introduced the Lefaucheaux-type gun to England, bought Hughes's patent and produced lever-cocking guns with exposed hammers.

By 1878, James Lang had developed his own design for a hammerless action gun cocked by barrel leverage. But when Anson and Deeley sued William Scott of Birmingham for infringing their patent, Lang — whose barrel-cocking gun, like Scott's was quite similar to the Anson and Deeley — went back to the Hughes design, this time in hammerless form. As with the Gibbs & Pitt and the Murcott systems, the Lang's under-lever both cocked the locks and retracted the bolts.

Handy though it was, even by today's standards, the under-lever couldn't compete with the Anson and Deeley system, and one by one, virtually every gun maker in England and the world came up with some version of an action cocked by leverage from the barrels. With that and with the widespread use of the Purdey bolt, the evolution of the double gun was, by the end of the 19th century, essentially complete.

Of the dozens of gun makers who, through the 19th and 20th centuries, contributed to the development and refinement of the sporting shotgun, at least three deserve special mention. Though they lack Purdey's or Holland's widespread fame, W&C Scott, John Dickson & Sons and John Wilkes characterize the quality and creativity of the British shotgun, past and present.

Ironically, W&C Scott, whose guns were

widely exported to the United States and the Crown colonies, was probably the most famous British gun-making firm in the world market during the 19th century. William Scott founded the firm in Birmingham in 1834, joined in partnership by his brother Charles in 1840. By 1858, William's two sons, William Middleditch and James Charles, had joined the company, which was renamed W&C Scott & Son.

William M. Scott became head of the firm in 1866, lending both a talent for gun design and a flair for marketing. When he retired in 1894, Scott guns were known worldwide. Of the 14 patents entered in his name from 1865 to 1884, the first proved most enduring. The Scott spindle was such an efficient means of connecting a gun's top lever to the Purdey-type underbolt that it is still used in Scott guns and in many built by other makers.

Other William M. Scott patents, some developed in collaboration with Thomas Baker, covered lock designs, cocking indicators, gas-vent systems, rib extensions and crossbolts, safeties and a compensating barrel lump in which looseness in a gun's hinge joint may be removed with a screwdriver.

In 1897, Scott merged with P. Webley & Son to form Webley & Scott, makers of military and sporting arms for more than 80 years and two world wars. During those years, Webley & Scott produced thousands of guns for their own trade and supplied thousands more, either fully or partially finished, to other makers. By the 1970s, the firm was the only supplier of gun barrels in all of the United Kingdom.

With all its success, Scott's shotgun market declined in the 1970s, and in 1979 the company announced the discontinuation of all double guns. Harris & Sheldon, the parent company that had purchased Webley & Scott in 1973,

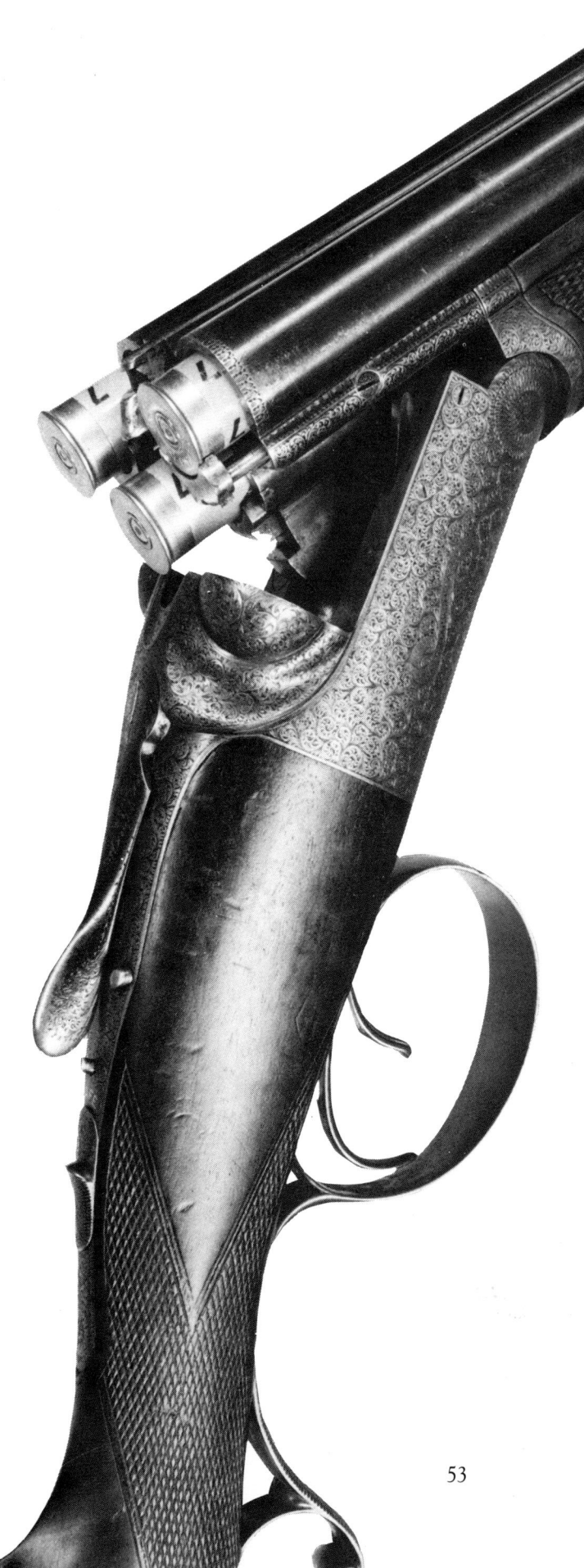

A very rare three-barreled John Dickson shotgun. Photo courtesy of Geoffrey Boothroyd.

revived the W&C Scott name and set up a new factory in Witton, dedicated solely to the production of fine shotguns. Today W&C Scott (Gunmakers) Ltd., is virtually the only gun maker in England that is not only surviving the turbulent world economy but actually expanding its markets around the world.

In an era when the trend in British gun making was toward building guns more alike, John Dickson's guns were unique. Though its history goes back to the first John Dickson's shop on Edinburgh's High Street in 1820, the company's fortunes were most firmly based on the "round action" guns developed in the 1880s, during the last few years of the third John Dickson's life.

Called the round action because of smoothly rounded contours, the Dickson gun represents a technical link between sidelock and boxlock designs. The lockwork is mounted on the trigger plate rather than in the frame or on sideplates. Its strong action lends itself to lightweight guns.

The round action proved versatile enough to accommodate designs other than the traditional side-by-side double gun. One of the earliest, patented in 1882, was the Dickson three-barrel — three tubes mounted side by side, fired by three triggers. A lever on the right-hand side opened the action. Though there was a flurry of interest in three and four-barrel guns during the 1880s (A. A. Thorn, owner of Charles Lancaster's old firm, patented three different four-barrel shotguns in 1881 and 1882), the fad was short-lived. Dickson's built only a dozen of their three-barrel guns.

In its latest application, in the 1930s, the round action, turned 90 degrees and fitted with a side lever, became the Dickson over/under. Once again, only a few were built.

Although the third John Dickson, who died in 1886, was the last of the founding family, the firm has survived. Through mergers over the years, Dickson's acquired ownership of such great Scottish gun-making companies as Mortimer, MacNaughton, Harkom, Alexander Henry and Alexander Martin. It is now known as Dickson & Martin.

John Wilkes is the oldest gun-making company in England which is still owned by the founding family. Since the first John Wilkes set up shop in London in the 1830s, the business has endured generation after generation in the classic mold. While other makers flourished and changed and died, Wilkes has remained much the same sort of small shop that men in tall hats or Edwardian broadcloth coats once visited.

Over its years, the firm has built everything from muzzle-loading pistols to military weapons to sporting rifles. Today at the Beak Street shop a handful of craftsmen, including the present John Wilkes, build only shotguns — slender, elegant pieces fitted and finished with a skill in every way equal to London's Best.

In the 1980s, a custom-built double shotgun is a megadollar anachronism. As John Wilkes puts it, the English gun trade is a 19th-century industry struggling to survive in the Space Age. Even the plainest mass-produced British gun these days will exceed the cost of most home computers; a custom-made gun costs more than most automobiles. The number of English gun makers dwindles year by year, along with the number of craftsmen sufficiently skilled to build a British gun. Still, a market remains. There seems to be enough well-heeled sportsmen and collectors in the world to support a lively trade in both new and used English guns of the modern era. Best of all, the old ones, artifacts of the greatest revolution in firearms history, are still around, offering a link between our world and a time not likely to come again. ♦

Winchester's
MAGIC DOUBLE

Winchester's handsome Model 21 is the last remaining tie to an era of fine double shotguns — the essence of American "built by hand" craftsmanship.

JERRY WARRINGTON

Resting there among the Parkers and the Purdeys, the Foxes and the Holland & Hollands, it had a subtle sort of inbred quality about it. Amidst a rack of double guns, the names of which read like a historical "Who's Who" of gunmakers, the small-framed side-by-side was a relative plain Jane in comparison to the more ornately engraved guns nearby. The barrels were sorely in need of bluing and the receiver had long since taken on the soft, well-worn patina of heavy use.

Turning to the man behind the counter, the young wingshooter's eyes asked a silent question. Permission granted, he gently lifted the 16-bore double from its allotted place and cradled it in his hands. The balance was good. No, better

than good; it was exceptional. The straight-gripped double gun came easily to the shoulder, bringing with it the heart-stopping flush of a grouse from some distant hillside stand of aspen and woodcock holding steadfast within inches of a setter's nose.

Bringing the gun to rest at his side, he rolled the briar-scarred buttstock and examined the gun more carefully. It was then he glimpsed the markings atop the barrels and grinned in recognition: 21. He should have known. It was as lucky a number for a shotgunner as for a blackjack player in a Las Vegas casino. For the gun fancier, however, it's the Winchester Model 21, one of the last of the "built-completely-by-hand" American double guns.

It was during the 1920s that the technical staff at the Winchester Repeating Arms Company started working on what would eventually become known as one of the finest double guns ever built in the United States. To insiders among gunning circles, it was a questionable venture at best, since the new shotgun would be forced to compete with such well-established favorites as the Parker, A. H. Fox, Ithaca and L. C. Smith side-by-sides. Still, in respect to an increasing trend toward conservation, Winchester moved ahead with the project while continuing to develop other designs through the foresight of Thomas Crossley Johnson, creator of the Model 12, and William Roehmer. By retaining its diversity, the company protected itself from a problem which was to plague a number of other gun firms — total commitment of financial resources to the double gun trade.

Another reason why many people questioned Winchester's interest in a double gun at that juncture was the financial status of both the company and the world. Following on the heels of World War I, monetary systems were in

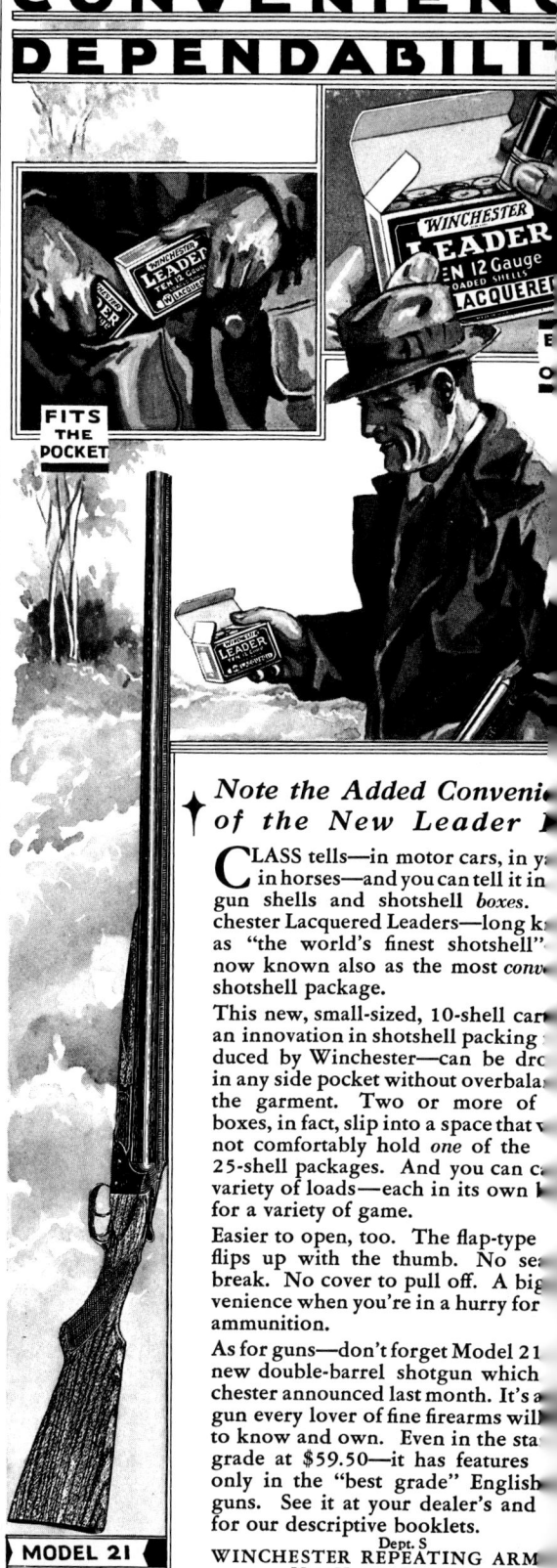

This 1930 advertisement for shotshells also boasts of the Model 21, which Winchester had recently introduced at $59.50.

complete disarray, with inflation totally out of control. In essence, it was the worst of times to become involved in what appeared to be an uncertain proposition.

Winchester's financial state was also uncertain. In response to an urgent demand for increased production for military contracts during World War I, Winchester made extensive additions to its manufacturing plant. Unfortunately, in its haste to meet the needs of the military, Winchester failed to make its new buildings efficient enough to be used after the war. Once the war was over, Winchester found itself with extensive building liabilities rather than assets.

To further complicate matters, Winchester was caught up in a series of fixed-price contracts, the result of their involvement in military production. Rising wage costs and overexpanded facilities forced them to seek alternative consumer markets — hardware products, roller skates and the like. Finally, the Winchester Repeating Arms Company went bankrupt, its stock falling from an all-time high of $2,500 a share to little, if any, value at all.

The Winchester Model 21, however, had progressed well during the years of financial instability. Despite the obvious signs of pending doom for the firm, the technical staff which included the Model 21's principal designer, George Lewis, continued its design work, at last introducing the side-by-side in 1930. It was a boxlock in nature, boasting a unique feature: its chrome molybdenum steel barrels were joined by a vertical dovetail method, adding additional strength and interlocking the entire assembly.

Collectively, the first Model 21s carried the standard pistol grip with double triggers, splinter fore-ends and both plain and selective ejectors. Barrel lengths ranged from 26 to 31 inches with a variety of choke combinations — improved cylinder/modified, modified/full, and full and full. Beyond that, the Model 21s were anything but standard. Though conceived as a top-quality production gun with broad-based appeal, the shotgun became the result of strict attention to detail. Close tolerances were required in the machining processes, and the side-by-sides eventually were assembled and fit by a small group of selected gunsmiths. Despite production-grade status, the Model 21s received a great deal of handwork, even in the early years.

In 1931 the Winchester Repeating Arms Company was purchased by the Western Cartridge Company. With the acquisition came John Olin, a son of the founder of Western Cartridge, and the man who would firmly establish the Model 21 as one of the best ever made by American craftsmen. His arrival at Winchester's New Haven complex heralded the beginning of a new era. Recognizing the Model 21 for its inherent qualities and bowing to his own admiration for fine shotguns, his first order of business was to increase promotional emphasis on the gun. The subsequent sales over the next few years proved Olin to be an excellent marketer.

During the next decades, few internal changes were made in the 21, with the exception of adding an optional single trigger which, along with selective ejectors, became a standard feature in the early 1950s. In fact, the mechanics of today's Model 21 are literally the same as they were at the gun's inception.

On the other hand, the external design and appearance saw a number of changes, including the addition of an optional ventilated rib. To accommodate the new ribs, two alternate frames were designed with higher ramps — one in 12-gauge, another in 16 and 20. In all,

there were four frames to choose from — the two for the standard-ribbed guns and those for the later models.

The most dramatic change in the Model 21's appearance originated from an experiment by Ernie Simmons, Sr., an exceptional gunsmith. At a national skeet championship in the late 1940s, he debuted a .410 version of the Model 21, made by mounting twin Winchester Model 37 .410 barrels to a 20-gauge Model 21 frame. Though aesthetically unattractive in comparison to other Model 21s, the .410 was unique, and Dave Carlson, Winchester's production supervisor, was told to make one.

Winchester's first Model 21 .410 was a poor handling gun with little cosmetic appeal. Because of the angle necessary to fit the smaller barrels to the frame, the side-by-side was improperly balanced with most of the weight favoring the buttstock. Fewer than 50 of the actual Model 21 .410s were ever sold.

Despite its lack of success, the .410 design contributed a modification that became a standard for the Model 21 in 1960. To counteract the balance problems which plagued the .410 versions, the actions were milled and lightened in various areas. To reduce the overall weight, the arrow-shaped side panels, which had long been a part of the 21, were removed. After extensive testing with the lighter-weight frames, Winchester's design personnel found that the side panels were indeed unnecessary and thus altered the action.

The frame alterations in 1960 coincided with the Model 21's transition from a "production-grade" gun to a strictly custom gun. Prior to that point, "custom" Model 21s had been available for about nine years along with the production guns. But once again rising costs spelled the end of the production phase, and in 1960 the

More wood is rejected as unsuitable for a Model 21 than is accepted. The grain of the wood is clearly revealed as the buttstock begins to take shape.

Winchester Custom Shop became the sole manufacturer of the custom-built Winchester Model 21.

Following reassignment to the Custom Shop, the Winchester Model 21 was offered in three distinctive grades — the Custom grade, Pigeon grade and the premier model, the Grand American. The major differences between the three revolved around the options available and their external appearance.

> *The Model 21 should gain a fair share of new followers in the years ahead. In the past, it has been privy to gun racks belonging to presidents, generals, and lesser-known wingshooters who enjoyed fine shotguns. Given the quality that has been, and will always be, the trademark of the Model 21, it will undoubtedly weather future decades.*

The Custom grade was the least expensive Model 21. The gun carried a hint of engraving and a checkering pattern which was relatively simple to execute, and the grade wood was somewhat less than top-of-the-line, full-figure walnut. In spite of the lesser costs, the Custom grade still featured the same hand craftsmanship and finishing touches of the two more elite models.

The Pigeon grade was very similar to the Grand American. The engraving patterns were identical, save that of the gold inlays most often featured on the Grand, and the checkering pattern was similar as well. Ultimately, it was the Pigeon grade's almost exact duplication of the Grand American in appearance and price tag that took it off the market toward the end of the 1970s.

Finally, there was the Model 21 Grand American, THE Winchester gun in terms of ornate master crafting. Each came with two sets of barrels and fore-ends. In addition to the finest engraving, it featured a variety of field scenes inlaid in gold. And when it came to wood selection, the Grand American knew few equals. The crotch figure walnut on almost every Grand American was exceptional, carrying a norm of 24-lines-to-the-inch checkering.

Things seemed to be going well for the Winchester Custom Shop and the Model 21 during the 1970s. Orders were reasonably steady, evidenced by an average wait of 12 to 18 months for a completed Model 21. However, in 1980 it appeared that the Model 21 would once again succumb to financial pressures. Olin/Winchester announced that a good share of their New Haven facility — including the Custom Shop — was for sale and that no additional orders would be taken on the Model 21. If the complex was not sold by December of 1981, the shop and the factory would shut down entirely.

The offers were few and far between, and usually much less than Olin was willing to accept. Insiders were puzzled by the lack of response and many feared the end of the Model 21's reign. Then, two former Winchester executives brought together a group of investors, and on July 20, 1981, the Winchester line — minus the Olin-retained Japanese guns and the Winchester ammunition — became the property of the U.S. Repeating Arms Company,

the newly-formed alliance. Shortly thereafter, the Custom Shop again began taking orders for the Winchester Model 21, guns to be crafted by the same hands that had made them in years past.

In today's lineup of Model 21s, there are once again three grades: the Standard Custom-Built, the Custom grade, and the Grand American. The Standard Custom-Built, at $7,500, is the least expensive of the three. The Custom grade retails for $10,260 and, as the price indicates, it features more elaborate custom work.

The Grand American remains atop the heap, boasting a price tag of $21,060. It offers all of the features of the Custom grade plus extensive engraving, more detailed checkering, an interchangeable set of barrels, and a custom oak and leather gun case. Unlike the other two which are available only in 12, 16, or 20-gauge, the Grand American is available in 12, 16, 20, 28 or .410. The cost of a grand American in gauges 28 or .410 is $32,400 — $37,800 if you want one set of 28-gauge barrels and one set of .410 barrels rather than two sets in a single gauge.

For avid collectors, there is one very special Grand American. Actually, it's a "One of Eight" series of small-bore editions with three sets of barrels in gauges 20, 28 and .410. The barrels are all 26 inches, and the features of the limited editions are almost too numerous to list. The set retails for $54,648. For more information concerning the sets or any Model 21, contact U.S. Repeating Arms Company, 275 Winchester Avenue, P.O. Box 30-300, New Haven, CT 06511, or call (203) 789-5000. (By the way, as this is being written, the company has placed a moratorium on all new orders because of the backlog that developed in 1987. It would be wise to check with the firm regarding tentative delivery dates, as well as the possibility of having any refurbishing work done on older Model 21 doubles.)

In regards to collectibility, the Model 21 has held its own on the marketplace. As is the rule with any of the older shotguns, the smaller the gauge, the more desirable the shotgun. The condition of the gun will influence the asking price as well, but not nearly as much as the grade itself (which saw many variations in the early years). Considering the prices, the Model 21 seems to be doing rather well for a shotgun that was introduced for $59.50.

The Model 21 should gain a fair share of new followers in the years ahead. In the past, it has been privy to gun racks belonging to presidents, generals, and lesser-known wingshooters who enjoyed fine shotguns. Given the quality that has been, and will always be, the trademark of the Model 21, it will undoubtedly weather future decades.

And why not? The Model 21 is, after all, one of the few remaining ties to an era in which fine double guns were the hallmark; the essence of American "built by hand" craftsmanship. ◆

In the 1960s the Model 21 made the transition from a production-grade gun to a strictly custom gun. Today it continues to be the result of handcrafted detail.

ANSLEY H. FOX
The Man, The Legacy

Brilliant, demanding, and always restless, Ansley H. Fox bounced from one manufacturing company to another, but never abandoned his pursuit of excellence and "the finest gun in the world."

MICHAEL McINTOSH

He was born on the cusp between two worlds, at a moment when history seemed about to pause, take stock of things and move on in some new direction. Behind lay the old feudal culture of the South, devastated by war and hammered by Reconstruction into a caricature of its conqueror. On his first birthday, the Unkpapa Sioux would meet briefly with Custer on a far-off Montana prairie. Further ahead lay the 20th century, the new America where technology would blossom, an America peering for the first time past her own borders, an America that soon would prosper and consume and grow restless. He would be restless along with it.

Ansley Hermon Fox was born on Friday, June 25, 1875, in his father's house in Decatur, Georgia. His father was Addison C. Fox,

The engraving style on most Fox shotguns is indication of whether it was built before or after 1914. But the CE (ejector) Grade, above, was always engraved with English-style scroll, making identification of vintage difficult. Photograph by Richard W. Smith

homeopathic physician; his mother, Louisa Ansley Fox. By 1880 Addison Fox had moved his family to Maryland, where four more children were born.

The family lived in Baltimore, on West Fayette Street, as early as 1892. There, Ansley Fox's liking for guns and shooting flourished. M. H. Wright, in a sketchy biography published in *Field and Stream* in 1908, says that from an early age Fox's marksmanship "began to attract attention among his neighbors," which could mean anything from real skill to a penchant for shooting lightning rods off rooftops. Perhaps it was a combination of both, for he would soon prove himself to be a brilliant shot.

Wright goes on to say that Fox's proficiency with a gun prompted him to earn a living as a market hunter for three years following the end of his public-school education. If this is true, he probably was moved more by the virtually endless shooting around Chesapeake Bay than by need for a livelihood. Market hunting was at best scarcely a subsistence, but the romance of it must have been wonderfully appealing to the adolescent son of a comfortably well-off and apparently somewhat indulgent family.

However casual his market gunning may have been, Ansley Fox soon discovered a game to take seriously — trap shooting. Here, too, were worlds in transition. The great days of live-pigeon shooting were coming to a close, as the clay target was about to change the nature of the game. Like the great professional tournament and match shooters of the turn of the century — Rolla Heikes, James A.R. Elliot, William R. Crosby, and the ilk — Ansley Fox was a pigeon shooter.

By the mid-1890s he had built some reputation around Baltimore as a pigeon shot, and on October 14, 1898, he won the Maryland Handicap, killing 24 of 25 pigeons at 29 yards rise. But he'd already had a taste of fame by then — from the gun he designed.

A correspondent to *The American Field* of March 7, 1896, describes a hammerless, break-action gun "that bids fair to be a great success."

It actually was a revised version of an earlier design. The first gun, described in a patent application filed when Ansley Fox was 17, never existed except on paper: the second one would change Ansley Fox's life. Joseph A. Geiji, a Baltimore gunsmith, built a prototype from Fox's plans, and the gun earned patent protection on June 30, 1896. Its design is similar to both Anson and Deeley's 1875 hammerless gun and to W.W. Greener's Facile Princeps action, invented shortly after.

The most prophetic features of this first Fox gun are its simplicity and its emphasis upon strength. The gun-making world, too, was in the midst of change. Breechloaders had been widely manufactured for little more than a generation. Though the hammerless action was well evolved, exposed-hammer guns were still much in evidence. The new nitrocellulose powders, more efficient and more powerful than black powder, were challenging the old concepts of metallurgy and gun design.

The 1898 Baltimore city directory lists Ansley Fox as president of the Fox Gun Company, 5 West German Street. This almost certainly was an office from which Fox contracted manufacture of his gun.

And in 1898, Ansley Fox took a wife, 16-year-old Fentress DeVere Keleher of North Carolina. Soon, but all too briefly, he had a son as well — James, born in December 1899, dead as an infant before the spring of 1902. The 1900 census, which mispells Ansley's name as "Ansel," shows them living at 421 North Carey Street.

Both the 1899 and 1900 Baltimore directories list Ansley Fox as secretary of the Fox Gun Company, located on Stockholm Street at the corner of Leadenhall. Burr Howard Richards was president and Burr Howard Richards, Jr., treasurer — clearly principal investors in what had become a manufacturing firm.

The company seems to have found some initial success. An item in the premier issue of *The Sporting Goods Dealer,* October 1899, says: "The Fox shot gun, invented and patented by Ansley Fox, one of the most promising young trap shots of this state, is being manufactured here now, but the trouble is that he cannot manufacture the guns fast enough to supply the demand." At least a handful of these guns still exist: twist-barreled, 12-gauge doubles stamped "Fox Gun Co. Balto. Md. U.S.A." on the left side of the frames.

There was a restlessness in Ansley Fox. Within a year, in a pattern that would repeat itself again and again, he changed his livelihood, striking off on a new adventure. In the spring of 1900, he became a professional shooter, engaged by Winchester Repeating Arms to represent its shotguns and ammunition.

What prompted his departure from the Fox Gun Company is as elusive as what moved him from one business to another through most of his life. The common assumption has been that he simply was an inept businessman, but it doesn't fit the facts. That he was able again and again to secure heavy investments in his ventures is hardly a sign of incompetence.

So what moved Ansley Fox? A powerful will, for one thing, probably supported by a healthy ego. It seems clear that Ansley Fox answered only to himself. He wanted to make high-quality items that bore his name, and nothing less would do. Sometimes, not even that would do, it seems, for he set out to dominate every market he entered, no matter what the product. And he was a man often out of sync with his time, sometimes ahead, sometimes behind. His markets often either were not ready for Ansley Fox or had already left him for something else.

Nothing in his life suggests a desperate seeking for financial security; he could, if he'd chosen, have found a lifetime career on any gun-manufacturer's design staff, but that was far too tame a prospect to interest Ansley Fox.

So with Ansley Fox gone from the scene, the Richardses regrouped in 1900 to form Baltimore Arms Company, makers of a shotgun that, while it owes something to Fox's 1896 design, was largely the invention of Frank Hollenbeck, a well-known designer of the time. By November 1904, Baltimore Arms was in receivership, its assets eventually dispersed.

Ansley H. Fox clearly sought to leave behind him something of quality, something that would endure.

Ansley Fox kept his residence in Baltimore through 1901, leaving the city frequently to follow the trap-shooting tournaments in the East and South, representing Winchester guns and DuPont powders and ammunition. He was genial, popular among the fraternity of professional trapsmen and with the public and sporting press as well.

He was a splendid shot, sometimes a brilliant one. He finished the 1900 Grand American Handicap tournament with a 96-percent average — the highest of anyone who completed every event — and yet failed to win the Grand itself. A month later, on May 15, he set a world record for clay-target doubles, breaking 98 targets out of 50 pair. He repeated his high-average performance at the 1901 Grand American and again failed to win the main event. In his two years of full-time shooting, Ansley Fox took part in dozens of clay-target and live-bird tournaments, but shooting never was meant to be a lasting career.

The Fox shotgun was adopted by the Savage Arms Company, but even that couldn't save it from becoming a casualty of World War II.

At the end of 1901 he resigned from Winchester, and by May the following year he was living in Philadelphia, gathering capital and machinery to build another shotgun.

In March 1904 notice appeared in trade journals that deliveries of new Fox guns would commence in July, at the rate of 15 guns per day. These were products of the Philadelphia Arms Company of North 18th Street and Courtland, in the Germantown section of the city. The company boasted a new, two-story brick building with 15,000 feet of floor space and a slate of officers headed by Ansley H. Fox, president.

The gun was a handsome piece. At first glance, you might mistake it for a V Grade Parker, with its rounded frame cheeks and dished sculpting around the hinge-pin. The bolting system, the central feature upon which Fox based his patent claim, comprises a rib-extension and top hook. All Philadelphia Arms guns were 12-gauges with Krupp steel barrels and coil mainsprings and top-latch springs.

Ansley and Fentress Fox lived ten blocks south of the factory, in a narrow row-house at 3534 Gratz, a small street that runs in fits and starts, a square or two at a time, north and south through Germantown. And soon he grew restless again.

On December 28, 1904, he sent a letter to the trade magazines, announcing his resignation from Philadelphia Arms and his intention to outfit a new factory that would produce a gun of his design. The official explanation for the split cited "business differences." There most likely were personality differences as well, for Ansley Fox was not amenable to counsel or compromise. At any rate, he found another set of investors among the wealthiest of Philadelphia businessmen and by April 1905 had filed articles of incorporation, citing capital of $100,000.

As promised, the news appeared in January 1906: The A.H. Fox Gun Company offered a new Ansley H. Fox hammerless shotgun, one to "compare favorably with the best of American or European makes on the market."

Ansley Fox, less modest, called it "The Finest Gun in the World."

Clearly, his new product was the culmination of Ansley Fox's notions on what a shotgun ought to be. It is simple, strong, and beautiful, with its three-part locks, piano-wire coil springs, and top-hook bolting. The first published announcements indicate that five grades were available, priced from $50 to $500, but only A, B, and C grades show up in the early advertising. Higher grades — D and F, which probably were special-order items in the first months — appeared in advertisements by 1907.

As Dan Lefever had a few years before, Ansley Fox found himself in the curious position of competing with his own guns in the marketplace. Through 1906 and early 1907, Fox Gun Company advertising included a line that read "Not Connected With the Philadelphia Arms Company." In November 1906 the company solved several problems at once by buying Philadelphia Arms, which not only erased a competitor, but also netted a virtually new, completely fitted factory. Patents and manufacturing rights naturally were part of the deal, which is why all but the earliest A.H. Fox Company Guns carry patents originally assigned to Philadelphia Arms.

Through 1908 and 1909 the company showed signs of prosperity. One-sentence news items in both February and August 1908 say that Ansley Fox was thinking of moving his gun factory to Havre de Grace, Maryland. In May 1909 another news item said the firm was considering expanding to accommodate a 60-percent increase in production.

Still, competition in the gun trade was fierce, and even at $50 the A Grade Fox was a relatively pricey item. In March 1910 Fox began courting the other end of the market with a new model, the Sterlingworth. At first, they were stamped simply "The Sterlingworth Company, Philadelphia, Pa.," obviously a trade name, since the gun was manufactured under patents owned by the Fox Company. Within a few months, the name was changed to Fox-Sterlingworth, and the gun appears in Fox advertising as the Model 1911.

Despite it all, the company was in trouble. Its advertising grew steadily more elaborate during 1910 and 1911, but there is a hint of desperation under the brave claims of flourishing sales. Finally, in April 1912, the trade was informed that Spencer K. Lewis had been appointed receiver for the A.H. Fox Gun Company. Reorganization, Lewis said, would be underway immediately. And Ansley Fox was gone again.

This time, at age 36, he was out of the gun business for good. By then, America was in the grips of a manufacturing frenzy, and though he clearly was disenchanted with the gun industry, Ansley Fox wasn't ready to abandon industry itself.

But change followed change. At some point, Wiliam Gerou, who worked as a machinist in the gun factory, had introduced his employer to his sister Ellen, and Ansley Fox discovered the great love of his life. He was married; she was not. One of the essential — and regrettable — differences between history and art is that history seldom records emotion. We'll never know what chemical spark was struck between Ellen Gerou and Ansley Fox, nor will we ever know what agonies passed between Ansley and Fentress Fox, but it ended in the Philadelphia Court of Common Pleas in the March term of 1913, when Fentress Fox petitioned for divorce.

Following the court hearing, Ansley Fox drops from sight for an entire year. It's possible — and my romantic bent would have it so — that he and Ellen lived together, forging the closeness that would endure for nearly 30 years, but that is scarcely more than wishful thinking, for there is no solid evidence to support it.

Still, their intentions were clear enough. On March 23, 1914, Judge J.M. Patterson issued a decree of divorce in the case of Fentress and Ansley Fox. Two days later, Ansley Fox and Ellen Gerou applied to the Clerk of the Orphans' Court of Philadelphia County for a license to marry. The following day, March 26, 1914, they were married at All Saints' Lutheran Church by Pastor F.A. Bowers.

They lived in what had been Ellen's house, at 244 Berkeley Street. By 1916 Ansley Fox was president of the Fox Pneumatic Shock-Absorber Company. He appears in the 1919 Philadelphia directory as president of the Fox Motor Car Company. His automobile was unveiled in January 1922 at the Hotel Commodore in New York City. True to form, it was beautiful, imaginative, and expensive — a two-seater coupe as trim and racy as a 20-gauge double gun. Its air-cooled engine housed six cylinders with aluminum pistons, overhead cams and valves, and developed 50 horsepower.

The Fox automobile made a good showing while it lasted. The 1923 model was a handsome sedan with a larger engine. Its impressive power and resistance to overheating made it a particular favorite among bootleggers. Unfortunately, not even they could buy enough $3,900 Foxes to keep the car on the road at the end of 1923.

Fox appears in the 1926 Philadelphia directory as president of the Fox Holding Company. It, too, was short-lived, and within a year he left Philadelphia.

As it turned out, he didn't go far. By 1930 Ansley Fox lived in Atlantic City, New Jersey, and was involved in a real-estate development corporation. In 1933 he moved to 214 Palermo Avenue in Pleasantville, on the western edge of the city.

Back in Philadelphia, the A.H. Fox Gun Company was not setting the shotgun world afire, but it was burning brightly enough. Within months after the receivership, new investors bought in, principally Edward H. Godshalk and his son Clarence. Their first decision was to expand.

To this time, Fox had built no guns in gauges other than 12. But by 1912, ammunition was efficient enough that smaller bores were becoming popular, and Fox introduced its first 16- and 20-gauge guns. The factory put out a small descriptive brochure to commemorate the event; it is subtitled "The Most Perfectly Proportioned Small-Gauge Guns Ever Built."

As with many of Fox's claims, there is some truth in it. Actually, 16- and 20-gauge Foxes are built on the same frame, scaled a bit closer to 16-gauge than to 20. But the standard Fox 12-gauge frame is smaller than that of nearly every other American maker, and the small-gauge frame shows exceptional harmony in the balance of proportions between frame and barrels. To keep the weight down, the company commissioned its own formula for steel, wonderfully stout stuff high in chromium alloy, trade-named Chromox.

Till Chromox appeared, Fox barrels were bored from Krupp steel blanks. Twelve-gauge guns were barreled with Krupp steel until World War I shut off the supply. Some small-gauge guns built between 1912 and 1918 have them as well, but at least as many were made, both barrels and frame, of Chromox. Following the war, all Fox guns were made of Chromox.

The Fox Company made minor changes in its engraving patterns over the years, but about

1910, the same time William H. Gough took charge of the engraving section, the style began to evolve from the intricate English scroll of the early days to the bold chiseling of later years. By about 1914 the transition was complete — except for the C Grade, which always would be engraved with graceful scroll.

And there was a new grade, the XE, between C and D in decoration. Catalogues of this period begin mentioning guns built to order. Single triggers, designed by Iowa gunsmith Joseph Kautzky, had been available on special order for a few years. In 1914 the company bought exclusive manufacturing rights, and the Fox-Kautzky single trigger appeared as a catalogue item.

World War I brought hard times. From 1916 until the Armistice, shotgun production slowed to a trickle while the company, scuffling for work, turned out rifle barrels, Very flare pistols, and parts for the Colt .45.

By 1919 the B Grade, now rarest of the lower-grade Foxes, had disappeared from the line, and a new single trap gun was introduced. One of the design projects interrupted by the war, the massive and deep-sided single trap isn't the most fetching specimen of its sort ever built, but it performed with the best. It was built in four grades — J, K, L, and M. The higher grades are heavily engraved and finely finished.

In 1922 yet another new gun appeared, the Super-Fox. Though the factory identified it as the HE Grade, the Super-Fox actually amounts to a separate model. In appearance and mechanics, it is identical to all other Foxes, but the Super was built on an oversized frame and fitted with thick-walled barrels carefully bored and regulated for dense patterns at long range.

The Super-Fox was made in both 12- and 20-gauges. The 20s weighed as much as 8¼

Ansley H. Fox appears in the 1919 Philadelphia directory as president of the Fox Motor Car Company. His automobile was unveiled in January 1922 at the Hotel Commodore in New York City. True to form, it was beautiful, imaginative, and expensive — a two-seater coupe as trim and racy as a 20-gauge double gun. Its air-cooled engine housed six cylinders with aluminum pistons, overhead cams and valves, and developed 50 horsepower.

pounds, and the 12s, built on what amounted to 10-gauge frames, as much as 9¾ pounds. Neither exist in great numbers; total production amounted to about 300 guns, only 59 of which were 20-gauges.

Standard chambering in the Super-Fox is 2¾ inches. Three-inch chambers, which had been available in 20-gauge Foxes since 1912, could be special-ordered in both gauges. Curiously, the standard chambering in most other Philadelphia-built Fox guns is 2⅝ inches. Some were bored with longer chambers, but not many. That's something to check out if you shoot a Fox.

No one knows what prompted Ansley Fox to switch from one business to another. But one thing is clear, he answered only to himself.

The company put out a special brochure in 1922, recounting in elaborate detail all of the Super-Fox's virtues — from a discussion of its specially overbored barrels (the bores actually are 11-gauge) to endless tables showing pellet counts from a multitude of loads. In a momentary excess of enthusiasm, the maker guaranteed full-choke patterns of 80 to 85 percent. The advertising section apparently didn't know that a shotgun is as fickle with its favors as a teenage beauty's heart, depending upon the cartridges you feed it. Before long, they stopped making promises the guns couldn't keep, which is why most Super-Foxes are stamped "Barrels Not Guaranteed" on the barrel flats. It refers to pattern density, not barrel quality. Super-Fox barrels are as stong as any ever built, and given a load it likes, the Super-Fox is super indeed.

Riding the wave of prosperity in the early 1920s, the Fox Company advertised guns even more highly decorated than the lavish FE Grade. From 1922 until 1930, the catalogues mention the GE Grade as available on special order. The price declined a bit in later years, but early on the GE was offered at $1,100. In the American market, only the $1,500 Parker Invincible cost more. Whether Fox ever actually built any GE Grade guns is problematic. None ever appeared as a catalogue illustration, and no factory record of one has come to light. In the early 1950s, Savage Arms vice president Herbert Stewart had two guns built that have since been advertised as GE Grades — a double and a single trap gun. But the Fox gun was no longer in factory production, and both of Stewart's guns were engraved in Germany, so one can pose a persuasive argument that neither is an authentic GE Grade Fox.

Even before the stock market collapsed, the A.H. Fox Gun Company was struggling to survive. By the end of 1929 the finest gun in the world was an economic albatross. In November that year, Savage Arms bought all manufacturing rights, machinery, and inventory and announced that production of the A.H. Fox Gun would be moved to the Savage factory at Utica, New York.

The Godshalks refitted the building with new machinery and set out to face the Great Depression as the Fox Products Company. For nearly 50 years, it turned out myriad light-industry items, from battery chargers and aircraft parts to locks, electrical outlets, and fishing reels. Fox Products Company remained in business until 1980.

Neither of the two great American shotguns that became foster children of the 1930s fared very well. Both Fox and Parker, taken over in

1934 by Remington Arms, were relics of a passing age, expensive to build and difficult to sell. Savage backed the Fox as a competitor in the target-shooting market with various trap and skeet guns — the Skeeter in 1931 and Trap Grade double in 1932, and various target-style Sterlingworths off and on for years. Yet another series of low-priced doubles, called the Fox Special or SP Grade, appeared in 1932.

By 1935, K and L grade single trap guns were no longer catalogue items; by 1937 only the J Grade remained. FE Grade was discontinued in 1940. The last Fox catalogue, issued in 1942, shows the various models of Sterlingworth and SP guns, graded doubles from A to D, and no single traps at all. The Fox Model B, which remained in production until 1987, was introduced in 1942. It shares nothing with the original Foxes except the name.

Like the Parker, Fox was a casualty of World War II. In the post-war world, the repeating gun was king, the double an antique. Savage assembled a few Foxes after the war, from parts machined long before. The last 12-gauge gun left the factory in 1945; the last of all the standard Foxes evidently was a 20-gauge SP Grade completed in December 1946.

Ansley Fox was past 70 then, retired and growing old. Ellen died of cancer at Temple University Hospital in Philadelphia on Christmas Day, 1942. Within a couple of years, Ansley Fox married Velma Shank, who was 28 years younger than he.

On August 9, 1948, less than two months after his 73rd birthday, a stroke left Ansley Fox partially paralyzed. The paralysis impeded the flow of his blood, and he developed pneumonia within the week. At 5:42 in the afternoon on Sunday, August 15, his heart finally stopped. He was buried four days later, next to Ellen at Harleigh Cemetery in Camden, New Jersey. An obituary in *The New York Times* on August 17 identifies him simply as inventory of the A.H. Fox shotgun.

Filtered through the gauze of history, the picture of the man is dim. He was a risk-taker, volatile, brilliant, demanding, even arrogant perhaps. Whatever internal tides moved Ansley Fox, he clearly sought to leave behind him something of quality, something that would endure. The search must have been painful at times, both for himself and for those whose lives touched his.

The world he knew changed and changed again; its currents often seemed to run against him. He must have felt some bittersweet satisfaction in seeing the company he founded continue to produce the gun he had created and named. But by the end of his life, even the gun was gone. How he felt about that, we can only guess. It's a pity that he never knew what remembrance history would hold for Ansley Fox.

GRANDAD'S G·u·n

Winchester's first pump-action gun, the Model 90 was no great beauty. And it probably knocked over as many tin ducks and pop-up targets as it did squirrels and rabbits. Yet the dependable little .22 was a shooting favorite for over 70 Years.

MICHAEL McINTOSH

My grandfather may not have been the worst wingshot I've ever known, but he was a strong contender. He never hunted in the years I knew him, but once in awhile he'd take a notion to thin out the barnyard pigeons; that exercise always resulted in a great uproar, but never in a single dead bird. His shotgunning career peaked one fall night when I was about eight years old. A clamor among the chickens got us all out of bed to see a red fox trotting back and forth by the henhouse door, ghostly in the moonlight. Grandad stepped out the back door, drew down on old Reynard with a load of high-brass sixes and blew up one of my grandmother's clothesline posts, which was about 30 degrees south of where the fox had been standing.

Part of the problem may have been his gun, which was for those days a typical farmer's shotgun: a single-shot 12 gauge of undistinguished pedigree with about eight inches of drop in the stock and all the balance of a potato fork.

But bad as his shotgunning was, Grandad was a wizard with a .22 rifle. Whatever was safe at his hands from a four-foot shot swarm (which was everything) was dead meat to a 40-grain bullet — feral cats, stock-marauding dogs, sparrows, starlings, an occasional crow or skunk. All it had to do was hold still for a moment. I once watched him drop a weasel emerging from under the henhouse; one offhand shot neatly through the shoulders at a distance from which the tiny animal looked to me no bigger than my thumb.

Whatever his shotgun contributed to the spectacular misery of his wingshooting, Grandad's .22 probably lent an equal degree of success to his riflery. I certainly was convinced of that at the time, and even now, when I know a bit more about guns than I did then, there is still something magical to me about a Winchester Model 90.

The love of small boys for grandfathers tends to cast a flattering light on things, but the fact is, the Model 90 earned a fair distinction all on

Photograph by Art Carter

its own. It was the first pump-action gun that Winchester built. It also was Winchester's most popular rimfire repeater, accounting for a production of about 849,000 guns in its original form, an almost identical number in a cheaper version called the Model 1906, and nearly another half-million in its final form as the Model 62.

The Model 90, like so many great guns, was the child of John Browning's splendid mind, and it came during a period when Browning's designs were helping to make Winchester preeminent among American gun-makers.

In 1885 Colt brought out a pump-action gun called the Lightning Magazine Rifle, built according to a patent secured by William H. Elliott in 1883. The Spencer pump shotgun had by then been on the market for a couple of years, and Winchester was beginning to see vast potential in the slide-action design. Winchester sent a Colt Lightning out to Browning in Utah and asked if he could design a better gun of the same sort. Indeed he could.

In December 1887 Browning filed application for a patent on a slide-action, exposed-hammer rifle that was simpler, stouter, and mechanically far more reliable than the Colt. The patent was issued as No. 385,238 on June 26 of the following year, and Winchester immediately bought manufacturing rights.

The Colt Lightning was chambered for .32-20, .38-40, .44-40, and .22 rimfire. In *The Winchester Book,* George Madis writes that Winchester originally intended to build Browning's gun for some big-bore centerfires, too. The action, with its massive breechbolt and twin locking lugs, clearly was strong enough to accommodate big cartridges and heavy loads. But, Madis says, the company decided that its lever-action guns had the big-bore repeater market cornered well enough and that the Browning gun would better fill the need for a new .22 rimfire.

All things considered, it was a wise decision. Pump-action shotguns ultimately all but took over the American arms industry, but no high-powered rifle of the same style has ever made much impression on the market. In popularity, pump .22s have fallen somewhere in between, and a big chunk of that territory belongs to John Browning's little gun. Besides, Winchester truly did need a new .22 repeater. Its first — and possibly the first repeater of any American make to be chambered for .22 rimfire — was the famous Model 1873 lever-action, built as a .22 from 1884 through 1904. It would hold a pocketful of .22 cartridges and lacked nothing in reliability, but the 73 simply didn't sell in any great numbers as a smallbore. About 19,500 were built as .22s, averaging fewer than 1,000 per year. By contrast, annual production of the Model 90 averaged more than 20 times greater.

First public notice of the rifle came in the November 1890 Winchester catalogue, and the

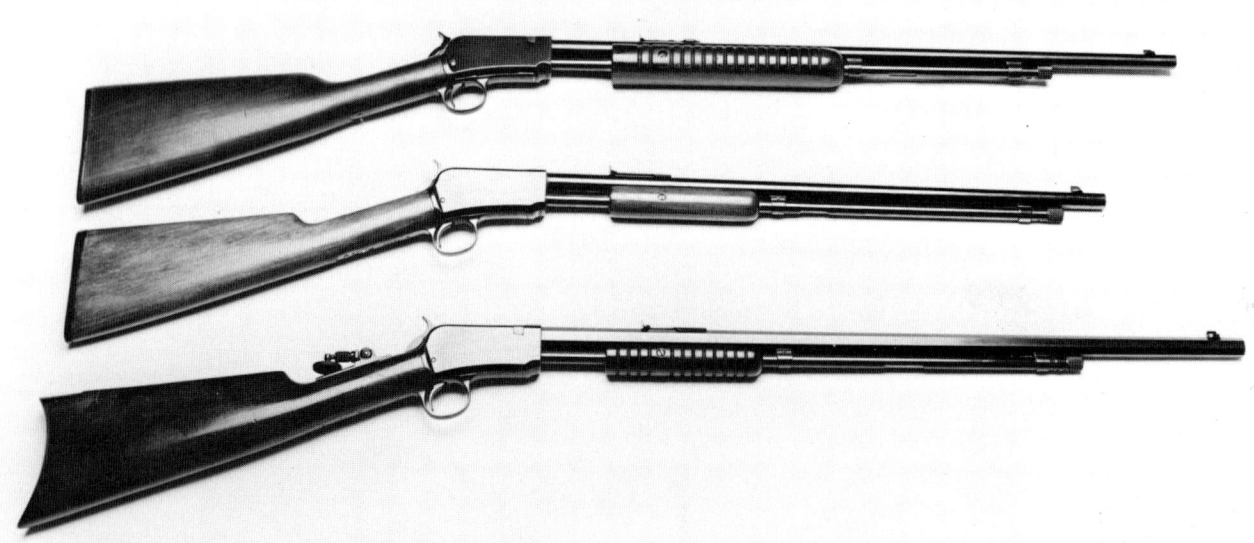

first guns were delivered to warehouse stock on December 1. According to the catalogue, only 24-inch octagonal barrels, plain triggers, and straight-grip stocks with curved butts were available. In later years, quite a few special-order Model 90s were made with pistol-grip stocks. A handful, undoubtedly also special-order guns, were fitted with set triggers.

The earliest Model 90s were solid-frame guns with color-case-hardened receivers, hammers, buttplates, and triggers. In the first 12,000 guns, both front and rear sights were dovetailed into the barrel; after that, the rear sight was held on by a screw, though a customer still could request dovetail fastening. About the same time, flatter, shotgun-type buttplates of steel or hard rubber became available on request.

Every Model 90 is cartridge-specific — the feeding mechanism and chamber are designed for one cartridge only, and the various .22 rimfires cannot be used interchangeably. The first guns were built for .22 Short, .22 Long, or for the .22 Winchester Rimfire, a cartridge invented specifically for the Model 90. (It was the first truly successful cartridge in what we now think of as the .22 Magnum rimfire class, far better than the old .22 Extra Long, developed about 1880.)

In December 1892 Winchester announced that the Model 90 would thereafter be built as a takedown gun, and the appearance of these represents phase two in its evolution. These second models, according to Madis, begin with serial number 15522 (though a few solid-frame guns have higher numbers); they are identical to the earlier model except that removing a large-headed screw from the left side of the receiver allows the gun to be broken down into two pieces for more convenient storage and transportation. When the gun comes apart, the receiver housing and breechbolt remain attached to the barrel-and-slide assembly. The rest of the lockwork stays with the butt section.

Around the turn of the century Winchester revised the 90's design slightly to create a third version. In these guns, the locking lugs, which in the first two models remained concealed within the frame, extend through the top of the receiver. Presumably this was done to strengthen the action, but since the older guns were more than adequate for .22 rimfire ammunition, the advantage may have been more academic than actual. In June 1906 a further revision resulted in two notches milled into the top of the receiver at the front. These provided additional locking surfaces at the front of the breech-bolt and probably strengthen the action somewhat.

Madis points out that second-model guns end with serial number 326615, but don't be surprised if you run across a Model 90 with a lower number that has exposed locking bolts. Manufacture of old and new models often overlapped until the supply of old-model parts was exhausted. This is true of lots of guns built by lots of makers.

In August 1901 Winchester discontinued its color-case treatment of several guns, the Model 90 included. From then on, 90s were blued overall. The transition period, during which both blued and case-colored guns were produced, covers serial numbers from about 97000 to about 105000. Even after blued receivers became standard production, case-hardening remained available on special order.

Until the end of World War I, .22 Short, .22 Long, and .22 WRF were the only standard chamberings for the Model 90, though a few were made up in .22 Long Rifle, probably on special order. Long Rifle chambering was put into regular production in 1919.

From top: Winchester Models 62, 1906, and 1890. The 90 was noted for its octagonal barrel and curved buttplate. The Model 1906 is a gallery rifle, which shoots only .22 Shorts.

For many years, Winchester offered custom engraving and finishing for all its guns, ranging in extent from minimal to elaborate. Factory engraved Model 90s are quite rare, but there were some. Rabbits and squirrels were popular for game scenes on receivers, though some 90s were embellished with running deer. (An apprentice in the Winchester Custom Shop also once engraved a pheasant on a Model 94 carbine. Logic does not always prevail.) Virtually all of the factory-engraved Model 90s I've seen were also fitted with figured-walnut stocks with pistol-grips and checkering. Lyman tang sights were popular accessories. A few guns were special-ordered with checkered fore-ends instead of the standard grooved slide handle; these apparently are quite rare, even among fancier guns.

Like so many great guns, the Model 90 was the brainchild of John Browning, whose splendid designs helped to make Winchester preeminent among America's early gun-makers.

The Model 90's handling and durability made it a particular favorite in shooting galleries, and Winchester turned out tens of thousands of them for the gallery trade. A number of these were specially marked on the receivers, usually with the name WINCHESTER in the company's classic staggered letters or sometimes with a large W filled with red enamel. Such marking was applied only on special order and, according to Madis, only if the Winchester sales representative in the customer's territory approved it. Gallery owners often misadjusted the sights, and Winchester was understandably loath to supply prominently marked guns to such people. The factory filled those orders only after the sales rep visited the gallery and pronounced it a square game. Sneaky tricks notwithstanding, the Model 90 probably knocked over as many tin ducks and pop-up targets as it did squirrels, rabbits, and assorted varmints.

Like all of Winchester's best guns, the Model 90 was a relatively pricey item to produce and therefore to sell. Courting the market for less expensive guns, Winchester designers took the Model 90 action, put on a 20-inch round barrel and a straight-hand buttstock with a shotgun-type butt and composition buttplate, and announced the gun in the January 1907 catalogue as the Model 1906. It was chambered for .22 Short only.

The .06's performance in the marketplace was at first less than overwhelming, probably because it offered no advantage over the Model 90 except in price. But in April 1908 factory engineers really did improve upon Browning's original design and changed the action so it would digest Short, Long, and Long Rifle cartridges interchangeably, making it the first Winchester rimfire repeater to do so. With that, sales took off like a scalded cat, and the Model 06 ultimately sold almost as many copies as the Model 90, which remained a one-cartridge gun.

A fancy version of the Model 06 was announced in 1918, but did not appear in Winchester's catalogue until 1924. It was called the Expert and had a pistol-grip stock, a specially shaped fore-end and, like the standard 06, a round 20-inch barrel. The Expert was available with a blued finish, with the receiver, trigger guard, and bolt trimmed in nickel, or with all exposed metal parts nickel-trimmed.

Both the 06 and the Model 90 sold well during the 1920s, and both took a nosedive with the onset of the Depression, along with every other sporting gun built in America. Both were taken out of production in 1932 and replaced by two new .22 repeaters that Winchester no doubt believed would fare better in a shrinking market. One was a hammerless, takedown pump called the Model 61, which was meant to compete with similar guns already available. The other was the Model 61, something of a hybrid version of the models 90 and 96.

The 62 was built on the more versatile Model 1906 action, able to handle all three standard rimfire cartridges. (Those built as gallery rifles, however, were made for .22 Short only.) The 23-inch round barrel owed something to both older models. Ironically, some of the design changes apparent in the 62's locking system turned out to be digressions rather than improvements, and the original Model 90 system was readopted for the Model 62 in 1938. The gun proved as popular as its predecessors and sold nearly a half-million before it was taken out of production in 1958. The Model 61, a Johnny-come-lately in the hammerless pump market long dominated by the excellent Remington Model 12 and its successors, made only a fair showing before it was discontinued in 1963.

Winchester continued to assemble Model 90s until 1941, when the supply of parts on hand finally ran out. The inventory apparently included more actions than barrels, because many 90s built in the later years have the same barrels as the Model 62.

To the dispassionate eye, the Model 90 is no great beauty. Even one that's gussied up with fancy wood and engraving shows little elegance or grace. But that's not to say that the gun lacks appeal. There is an old-fashioned quality about its spare profile that is handsome in the way of the Model 97 shotgun or the Winchester lever-actions or the Colt Peacemaker. The Model 90 is not a difficult gun to appreciate, but it doesn't, of itself, raise your blood pressure as some other guns do. I suspect that those who are truly fond of it love it more by association than by aesthetics. I know I do. Grandad's wasn't the first .22 I ever fired nor the last nor even the best. But it's the one I remember best, the one nearest my heart.♦

BROWNING'S
Last Credential

From the mind of John M. Browning, America's greatest gun inventor, came the Superposed, an over-and-under shotgun that has stood the test of time as a true classic.

TED SEFING

The Browning Superposed is a paradox. It is the culmination of an awesome inventive power, the last gun designed by a man described as "the greatest gun inventor the world has known." An art form among guns, the Superposed has often been called John M. Browning's masterpiece. Few will deny that the Superposed is a true classic — one of the most revered and desirable of all guns.

Yet the Superposed, in actuality, was more a design than an invention. Over and under design was not innovative in the 1920s when Browning, a gun inventor from Ogden, Utah, began to shape his Superposed concepts into walnut and steel. Over and under design was at least a century old and actually preceded the side by side. (Lower cost of manufacture and the ability to achieve compact action design gave almost total dominance to the side by side during the 19th century.) At least a decade before Browning's work on the Superposed began, such English firms as Beesley, Boss, Lang, Westley, Richards and Woodward were producing highly-esteemed over and unders.

The Browning Superposed really contained very few novel ideas. Most of the problems of two barrel design had already been solved by the great side by side makers. Under locking bolt design reached perfection by 1867 with James Purdey's double bolt. Hammerless boxlock design was worked out by Anson and Deeley in 1875. Automatic, selective ejection appeared by 1886 in the form of the Deeley ejection system, and the reliable Southgate ejection system was patented by Beesley in 1893. Also many designs for single selective triggers were in evidence by the turn of the century. Even the idea of incorporating recoil-affected, inertia pieces in the trigger mechanism to prevent doubling was not new.

Why then did the Browning Superposed become the world's most successful over and under shotgun? Why did John M. Browning turn his attention from repeating operation to a fixed action design in his last years? Why did the Browning over and under conquer the American market, when another respected over and under, the Remington Model 32, faded away during World War II?

There is no single answer. The answer is conglomerate; a mix of timing, cost, aesthetic appeal, forearm and trigger design, American prejudice and even conservation legislation. And not the least in significance is the indisputable fact that John M. Browning was indeed a master, and the Superposed certainly has great appeal as the last and most experienced credential of a genius firearms inventor. The feeling one gets from owning one is not unlike owning a painting by one of the great masters.

Browning perfected practically every breech-loading action type known in his day and today. He completely dominated the field of semi-automatic and automatic arms operation, holding complete mastery over the principles of short and long recoil operation, blow-back operation and gas operation. He sold many of his patent rights to the big arms companies in the East, so many of his famous inventions went afield under the name Winchester, Remington, Savage, Stevens and Colt. Starting with his Model 1900 automatic pistol in 1899, his company began a manufacturing relationship with Fabrique Nationale in Belgium that lasts to this day.

By age 45, he had startled the world with his remarkable inventive versatility. His credentials included the Winchester Models 85, 86, 87, 90, 92, 93, 94, 97 and 35 other designs covering single-shot falling block action, lever action, and pump action. One would expect his experience with these manual actions to precede his development of the more complicated, challenging automatic actions. But not so.

At 35, three years before he invented the famous Model 94 Winchester, he harnessed gas operation in an experimental machine gun that fired 600 rounds per minute. Concurrent with his work with manual action, he manifested semi-automatic and fully automatic action in such guns as the Colt "Peacemaker," Model 95 machine gun, numerous Colt .38 automatic pistols on the short recoil and blow back systems, and the Model 8 Remington recoil operated rifle. By age 44, he had completed his famous Automatic-5 shotgun and one year later, he gave the U.S. military the .30-caliber water-cooled machine gun.

All of his designs developed after age 45 were modifications and improvements to action types he had already created. Except one. That exception is the Superposed.

Browning began work on the Superposed in 1922, when he was 67. Up to this point, his consuming interest had been repeating operation. This is easy to understand,

The Four Bs, Ogden, Utah's top live-bird team: Left to right, G.L. Becker, John M. Browning, A.P. Bigelow, and Matthew S. Browning.

considering the time and his origin. By the mid-19th century, the ideas of cartridges and breech-loading were hard, metallic realities waiting for him to fashion into reliable repeating operation. Certainly, life on the Utah frontier asked for more than one shot.

What, then, was the catalyst that suddenly shifted Browning's interest to two-barrel shotgun design? Was it the popularity of side by side shotguns, which were highly esteemed by European, English and also American sportsmen of the early 20th century? No. Interestingly, the catalyst was the conservation movement.

Those were the days when America's conservation conscience was greatly troubled. And rightfully so. Game populations had been pressed into a tight corner by market hunting and rapid agricultural and industrial expansion that was decimating habitat. There was a great hue and cry for game legislation to protect this heritage. One of the targets of this legislation was the repeating firearm. The migratory bird law limiting shotguns to three shots was enacted in 1913. John M. Browning saw the possibility of severe restrictions being imposed on repeating firearms. One day he told his son he would design "the last firearm that could be legislated out of existence."

Along with his ability to see complicated mechanical functions, Browning had 20/20 marketing vision. He could see how pending legislation might pose a marketing threat, but, simultaneously, he could see that consumer need or desire would shield such a threat.

In a scrapbook kept by Gus L. Becker, one of his shooting cronies, an undated entry reads as follows:

"J. M. Browning called us together one day around the drafting table in the partitioned corner of his model-making shop. The table was covered with pencil sketches of details of arms mechanisms. Laid across one end of it were four shotguns of entirely different types, each the latest word in its class. He spoke about as follows:

" 'The Great American Gun Rack is pretty well filled, but there is a conspicuous gap among the shotguns. I have had an eye on that gap for a long time, and now that there are no urgent military jobs on hand, I am going to have a try at filling it. As a matter of fact, I have the gun pretty well worked out in my mind, and I have

told the shop to get ready to start on the model at once. Probably there will be several models before the gun comes just as I want it, for this is not to be just a good gun. It must combine in a harmonious whole all the requirements I have in mind — and some others I'll think of as I go along...' "

The key word in his quote is "American." For the "gap" was very real in America. While over and under design was not new to the world, it was new to Americans. The timing couldn't have been better.

The elite shotgun of the day was the side by side. But its design had reached a finality. During its evolution, when a new feature such as automatic ejection was offered, all previous extractor guns were to a degree outmoded. So shooters had a great tendency to buy new, improved models. There were really no improvements that could be made to create new demand, new market growth. Except one, perhaps. Rotate the barrels to over and under configuration.

"The Great American Gun Rack is pretty well filled, but there is a conspicuous gap among the shotguns ... I am going to have a try at filling it."

John M. Browning speculated that American shooters were ready for this "innovation." He was right: they welcomed it enthusiastically, for they already had a strong bias for single sighting planes.

In America, hunting hadn't polarized into the upper classes. Hunting was a great American privilege for everyone. A gun was a common household possession as well as a practical necessity, not an expensive luxury. Consequently, the guns favored by Americans were the more inexpensive, practical, single barrel firearms. Rifles were in great favor because of the seemingly unending resource of big game. There was a broad market for all categories of guns from the less costly single shot pieces to the modestly priced repeaters and even the expensive hand worked double shotguns.

Browning keenly foresaw the practical appeal a single sighting plane on a two barrel shotgun would have for Americans. He comments on this through Becker's scrapbook saying, "If you sight with a yard stick, say an inch wide and a quarter thick, you would turn up an edge. You couldn't sight with any accuracy along the broad surface. Vision would flood all over it and spill off. The side depth throws the edge into relief."

Indeed, there was a gap in the Great American Gun Rack. To Americans, over and under shotgun design was fresh and novel. In fact, there was at least one unique feature the Superposed had which the English and Continental over and unders lacked.

That feature was a sliding forearm, which the shooter did not detach to dismount the barrels. Upon reading Browning's Superposed patents, it is apparent that he felt the sliding forearm was a very good idea. (I believe, to this day, it remains unique to Superposed design.)

It allowed him to design a very compact and trim forearm, compared to the forearms on most English over and under designs. The inner, upper walls of the forearm could be contoured to fit snugly in the concave depression between the barrels, since there was no necessity to provide the clearance to pivot the forearm away from the barrels. The English makers solved this forearm clearance problem in two ways. The first

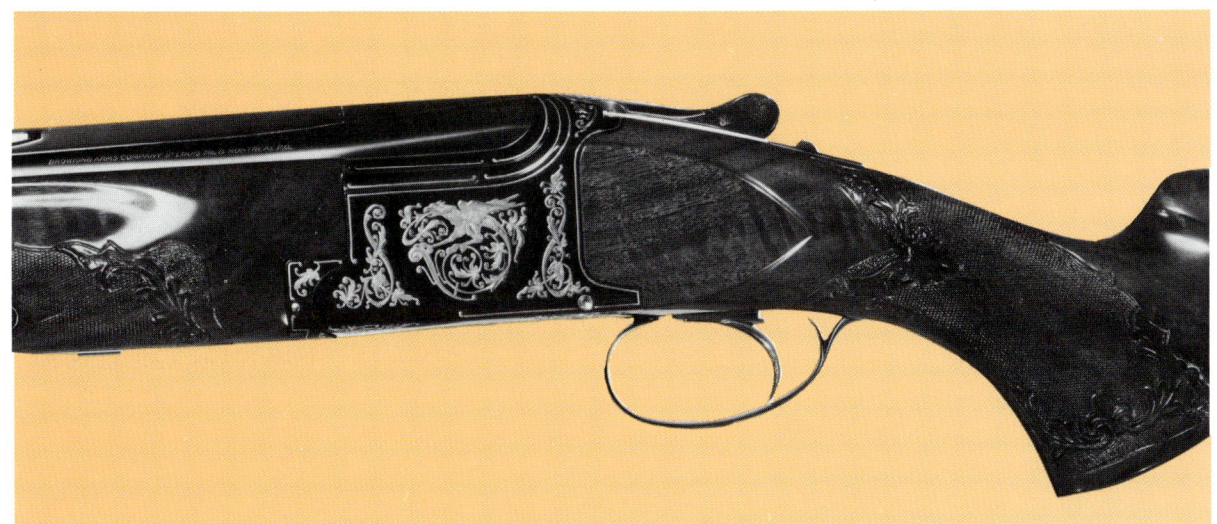

was to construct an expensive, three piece forearm, wherein the upper walls were separate pieces screwed into the barrel side ribs. The second solution was to design a shallow forearm which did not flare into the concaves. The former solution was costly. The latter solution was not as aesthetically appealing.

There are other benefits to the sliding forearm design. There is no chance to drop or possibly lose the forearm. And there is no tendency to squeeze the forearm too hard during disassembly and split the wood.

Another, less apparent advantage is added security to the barrels when mounted. The forward barrel lug is hooked onto the action hinge pin and the barrels rotated to the closed position. Now the cocking lever lifter which is solidly pinned to the forearm bracket is held upward and slid rearward with the forearm into recesses in the forward barrel lug. Thus, the hinge pin is completely encircled with steel when the forearm is latched rearward into position.

Yet another feature which contributed greatly to the Superposed's success was an inertia trigger design *which did not double*. An inertia trigger was not a new idea. (Major Gerald Burrard, in his classic work, *The Modern Shotgun*, points out that there were probably a hundred single trigger designs by 1910. Various delay mechanisms, including inertia weights, were built into these triggers to eliminate the great curse of single triggers — doubling, the near simultaneous firing of both barrels at the same time.) But a foolproof single trigger that didn't double was indeed unique, and the inertia trigger which John M. Browning's son, Val A. Browning designed was highly regarded as the trigger that couldn't double.

Doubling is actually the sequential firing of both barrels caused by a phenomenon known well to gunmakers — involuntary pull. Involuntary pull is a reflexive second pull of the trigger which occurs in an infinitesimal fraction of a second after the intended pull. Like a shoulder flinch, the shooter is unaware of this pull, and if it causes the second barrel to discharge, the firing sequence is so rapid the shooter thinks both barrels go off at the same time.

The Superposed has always been a favored canvas of the engraver's art.

Early single trigger designers knew little about involuntary pull. Many felt involuntary pull occurred after rearward recoil and was caused by the gun's rebounding forward from the shoulder into the finger. In actuality, involuntary pull almost always occurs during recoil, while the gun is moving rearward.

Again, Browning knew the American shooter. He knew Americans would prefer a single trigger instead of double triggers. He knew Americans preferred a pistol grip instead of a straight grip and that the angle of a grip would cause problems with double triggers. On a straight grip the fingers lay out of range of the recoiling trigger guard. With a pistol grip the fingers would frequently bruise on the trigger guard, especially when reaching for the forward trigger. When firing the rear trigger, the trigger finger would often bruise on the forward trigger. Too, there was the problem of gloves. Double triggers are not very manageable with gloved fingers.

Browning's patents clearly show that his design idea included a single trigger. Yet any Superposed historian knows that the first guns produced in 1931 bore double triggers and later twin single triggers. The single selective trigger did not appear in the Browning line until 1939, and for a very good reason.

Browning died in 1926 before the trigger was perfected. A master at utilizing the forces of recoil to operate his mechanisms, Browning was pursuing inertial movement to lock the trigger for the involuntary pull, but he hadn't completely solved the problem.

After Browning's death, his nephew, Marriner Browning, made some modifications to the trigger, but the prototype nearly toppled Marriner's cousin, John Browning, out of a duck boat. John Browning was president of Browning

Only a remnant issue of the Belgian Superposed remains in the Browning line today. That grim reaper of all that is sacred and nostalgic in fine, old guns — severely escalating manufacturing cost — has been dogging the Superposed for more than a decade.

Arms Co. at that time, and double triggers were scheduled for first production in 1931.

The problem was solved when Val Browning, son of the inventor and brother of John Browning, perfected the inertia trigger. Initially, Val incorporated twin single triggers on the gun. These differed from double triggers in that the shooter did not have to shift his finger to a second trigger to fire the second barrel. A second pull on the same trigger fired the second barrel. The firing order of the barrels was determined by selecting a trigger.

The twin single trigger did not double. It retained the fastest method of barrel selection which endeared shooters to a double trigger gun, and it went one step further offering the fastest second shot. Superposeds with twin single triggers were offered until 1940.

By 1939 Val Browning's single selective trigger appeared on the Superposed. It worked admirably, employing an inertia block to delay engagement of the second sear until recoil and gun rebound subsided. Barrel selection was accomplished by the position of the safety set by the shooter prior to shooting. Val Browning's single selective trigger contributed significantly to the success of his father's shotgun.

However the single selective trigger didn't appear till nearly a decade after the Superposed was introduced. During that first decade the Superposed did have one serious, high-quality competitor in America: the Remington Model 32. The Model 32 was announced in 1932 just one year after the Superposed debut. It was a French design and employed an over-locking bolt that bolted the barrels down from the top. This allowed a shallow receiver depth, admired on the English over and under designs. The barrels did not employ side ribs, and like the Superposed, early production had double triggers.

While the Model 32 was a top-quality gun, it did not survive World War II. The year 1942 is accepted as the year of its demise, and its obituary was written by escalating manufacturing costs. In 1932 a Remington 32 cost $20 more than the Browning, a time when $20 meant something. Estimated production runs were approximately 500 guns a year, definitely not enough production to amortize expensive tooling costs. Superposed production was approximately 2,000 guns per year, having reached 17,000 guns by the year 1939, when the war interrupted production.

The Superposed, made by Fabrique Nationale in Belgium, had one other advantage over the Model 32. It was the first over and under brought to America, beating the Remington by one year. Being first does count toward a stronger impression if the product lives up to its news value.

With its pleasing receiver surfaces, the Superposed has always been a favored canvas of the engraver's art. Over the years many famous engravers, both from Browning's staff as well as independents, have paid tribute to the design of Superposed with their skills. One would be hard-pressed to name another gun model which has been more frequently chosen for engraving.

High grade Superposeds first appeared in the production line about 1939. During that year, three high grades were offered. The first was called the Pigeon grade, featuring fine line engraving of pigeons in a delicate scroll background. The second grade was the Diana, and like the Pigeon, it had a greyed steel receiver with deep relief engraving featuring deer and boar. The Midas offered gold inlaid pigeons on a blued receiver.

In 1940 the war brought a halt to all Superposed imports, and it wasn't until 1948 that shipping resumed. For the first year only

Grade I 12-gauge models appeared. In 1949 the 20-gauge was offered for the first time. In 1950 Magnum 12-gauge chambering appeared, and high grades were again announced. The Pigeon, Diana and Midas designations were dropped; the high grades were now simply called Grade II, III, IV and V. The Grade II was similar to the former Pigeon grade. Grades III, IV, and V all featured greyed receivers, with pheasants and fighting cocks on Grade III; dogs, pheasants, and foxes on Grade IV; and ducks and pheasants on Grade V. In 1957 a Grade VI was added similar to the old Midas style, adorned with gold inlaid game birds on a blued receiver.

During the 1960s the old name designations were revitalized for the high grade, hand engraved models. (The smaller gauges, 28 and .410 models, were also announced in the early part of this decade.) The familiar Pigeon grade was offered. A new Pointer grade with fine-line engraving of bird dogs on greyed steel, ranked immediately above the Pigeon Grade. The former Grade V (deep relief pheasants and ducks on greyed background) was renamed the Diana Grade. The former Grade VI became the Midas Grade.

Only a remnant issue of the Belgian Superposed remains in the Browning line today. That grim reaper of all that is sacred and nostalgic in fine, old guns — severely escalating manufacturing cost — has been dogging the Superposed for more than a decade.

At first the symptoms were small and innocent. Such as the rounded, semi-pistol grip which was replaced by the "improved, modern" full pistol grip in the late 1960s. This happened about the same time the long, graceful trigger guard tang was replaced by a stubby, foreshortened tang. (The long tang has been reissued on today's Presentation Grades, but prices start at $4,500.) The semi-pistol grip and long tang were more time-consuming to hand finish and fit.

The beautifully hand engraved Pigeon, Pointer, Diana, and Midas grades were phased out by 1976 to be replaced by the Presentation Series — engraving motifs that begin with a chemically etched pattern that is touched up with hand engraving. Etching appeals only to the eye, not the heart.

Alas, the Grade I Superposed, which sold for $79.80 in 1931 and which still sold for a reasonable $375 in 1965, escalated to $1,100 just 10 years later in 1976. It was discontinued in that year. This year the Browning Company has reintroduced the Grade I Belgian Superposed in limited quantities at a suggested retail of $1,995.

Whenever hunters talk vintage doubles, the controversy over stack barrels versus "lazy eight" barrels is sure to come up. Each faction marshals forth its prejudices, the side by side connoisseurs claiming the advantage of the narrow sighting plane is more myth than reality. They tout their gun as less susceptible to cant, more compact, lighter in weight, with more beauty and symmetry of line.

I have always felt that the side by side, entrenched in tradition, carried a little more snob appeal and that the Superposed is, in fact, more practical, better balanced and stronger in design than even a back action, sidelock side by side.

The well defined sighting plane certainly is no disadvantage. The deep forearm does give good control and keeps the forearm hand lower, in alignment with the grip hand. The first shot fired from the lower barrel does not disturb aim as greatly as the first shot from a side by side. I have always felt that you can get on another

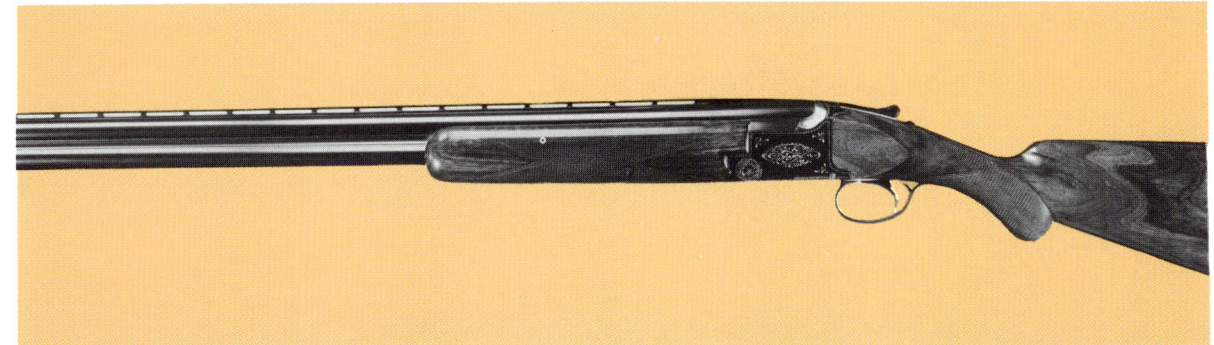

bird quicker with your second barrel.

Unlike most side by sides, and some other over and unders, the Superposed has a full length hinge pin which spans the width of the knuckle. Side by sides usually have dummy hinge pin heads on each side of the action because of the cuts in the knuckle which must carry the cocking levers. You also can't escape the great strength added by the girder-like walls of the receiver. Years of shooting have been known to crack side by side actions where the bar of the action intersects the standing breech. The deep side walls on over and under receivers add tremendous resistance to the downward stresses on the action table. Of course, these walls add a slight extra weight to over and under design, but the weight is added in the perfect place. The weight is distributed between the hands with the balance point exactly where it should be, very close to the hinge pin, varying slightly because of barrel length, wood density and other differences in individual guns.

No less an authority than Major Burrard agrees that over and unders offer better balance. Contrasting the balance of over and unders against that of a side by side he says... "there is usually a slightly greater concentration of weight in the middle part of the gun. The effect of this concentration is that there is less inertia in the ends and so the gun is easier to turn about its middle."

In my high school days our house had wall to wall linoleum. Our table cloth was red and white checkered oil cloth decorated liberally with my dad's cigarette burns. My mother cooked our game on a coal stove.

In those days of the mid-50s, a Grade I Superposed cost about $260. But linoleum floors or not, I ached for one.

I had never handled one, but I had seen its delicious, poised look. It looked like it knew what it was supposed to do. I had seen the way it could "turn about its middle." One shot was all it took to sell me. And I didn't fire the shot. My friend's uncle did. I'll never forget the grouse he swatted with a "Super."

The bird was a hard left, a blur, barnstorming out of a Pennsylvania thicket, not safe from my angle. So I just watched.

Uncle Bob's gun described a short arc, but it was an incredibly fast movement — almost a poke but with follow through added. Like the quick stroke you'd use to catch a fly in midair. The whirring wings folded in the midst of a single, ringing blast. The Superposed-smitten grouse crashed into a thicket but never felt the sting of brush. Someday, I'd have a gun that could do that.♦

Browning knew Americans preferred a pistol grip and that it would cause problems with a double trigger. His design idea for the Superposed included a single trigger.

American
CLASSIC RIFLES
The World's Finest

Made in America by a relatively small group of craftsmen, it's the best of its type ever produced, and you can have one made just for you. It's the classic hunting rifle, and it may not cost as much as you think.

ART CARTER

Wouldn't it be wonderful to someday own the very finest of a particular item that had ever been produced? Especially if that item happened to be a handsome gun made to your order? To hold such a masterpiece and gaze upon its beautifully figured wood, richly blued metal and flowing lines would be a dream come true.

Can't happen, you say. Those guns are made in England or Italy and cost as much as a house on the lake complete with a sailboat. Besides, even if you were a Baron or Duke or something and could afford such a prize, it takes four or five years to get one.

Well, believe it or not there is a type of gun, made in America by a relatively small group of craftsmen, that is the best of its type ever produced. And although not exactly inexpensive, a lot of everyday folks can afford one. In fact, you can have one made just for you. In many cases a gun ordered during the fall will be ready by the next hunting season. Sometimes it might take a bit longer but it's worth the waiting.

I'm talking about the custom-made American classic hunting rifle. These elegant creations are being sculpted by individual craftsmen all across the country. Whether they come from basements, garages or modern workshops, these rifles are the very best in the world. The rifles of no other country approach their excellence of line or fine craftsmanship. They are simply unchallenged.

Of course this type of rifle isn't for everybody. Called classic, they are just that. The stock is of the 'classic' style with subdued metalwork to add to its overall elegance. If you prefer the so-called California style then you are out of luck. Probably about 90 to 95 percent of

A pre-64 Model 70 Winchester classically stocked in Bastogne walnut by Jere Eggleston of Columbia, South Carolina. This .270W caliber has seen extensive use in Africa and the Northwest Territories. Photographs by Art Carter

all custom rifles made in the U.S. today are of the classic persuasion.

When a rifle is referred to as classic or California, this generally means stock style, but the rest of the gun is also important. California stocks usually have a shiny epoxy finish, slanted fore-end and flared grip cap of contrasting wood set off by white line spacers, high Monte Carlo comb with large cheekpiece, fancy skipline checkering, and a squared boxish fore-end. Some of the most radical examples may have a rollover cheekpiece that looks like surfs up at Wakiki or ivory and exotic wood inlays of everything from leaping tigers to gyrating maidens. With all of this goes a high polish on all the metal work. However, this has several basic flaws. Shiny epoxy finishes and super high-gloss blueing are just great for letting a deer or elk know exactly where you are. Rakishly styled Monte Carlo butt stocks are about the best way to insure maximum recoil. Sadly, the biggest flaw of all for many is a matter of taste.

Some sportsmen just don't care for all that flash and sparkle and instead love the subdued, elegant simplicity of the classic style. It is a graceful combination of flowing curves and straight lines blended into an efficient and beautiful package. Basically, the classic style stock has a time-consuming hand-rubbed oil finish that is in the wood, not on it. The butt stock has a high thick comb without a Monte Carlo. If there is a cheekpiece, it is a graceful continuation of stock lines. A comfortable, rounded fore-end will have beautifully executed checkering from 20 to 28 lines per inch or smaller, with 24 lines per inch average.

Checkering patterns will be wrap-around *fleur-de-lis* or multi-point patterns that require great skill to create. The contrasting fore-end tip, if one is used, will be of horn or ebony at a 90-degree angle with no white spacer. All blued metal will have a low-lustre rust or bead-blast blueing. Overall, the work must blend the rifle into a practical gun that won't spook game, with the elegance that only superior hand-craftsmanship can provide.

When you commission a riflemaker or stockmaker to build you a gun, the finished product will be individually yours. Like the 'bespoke' double-barreled shotguns of England, you can choose the final appearance of the gun so long as you stay within the parameters of your stockmaker's style. It will be your choice of action, wood, checkering, engraving, fore-end treatment, butt plate, grip cap, barrel, cheekpiece, and caliber that makes the gun just for you. The rifle can be super-light for mountain use or beefed up for one of the mighty magnum cartridges. It may be completely the work of one man or might be the work of several artisans. A stockmaker, metalsmith, blueing expert, and engraver may all contribute to the finished piece. However it is accomplished, the owner will possess a rifle that is uniquely his and not a copy of anything else.

All of this sounds great, so how do you get started and how far do you have to reach into your jeans to afford one of these guns? First you must choose an action. By far the three most popular actions are the pre-'64 Model 70 Winchester, reworked Model 98 military Mausers, and the Ruger No. 1. The first two are bolt actions and the last one is a single shot.

Both of the bolt actions are fairly expensive. A pre-'64 Model 70 will start at about $500 and go up depending on caliber. A good military Mauser action will cost less than $100 but at least $1,000 to $1,500 worth of metalsmithing must be done before it can be fitted into a custom rifle. A Ruger No. 1 barreled action

is about $225 up to about $350 for one of the rarer models.

The second largest expense (or maybe the largest) is the wood from which the stock will be fashioned. Most custom rifles will have some type of walnut as a stock wood. A few stocks are cut from such woods as maple, myrtle or cherry but these represent a small percentage.

Of the walnuts, the most popular is French, also called 'royal' walnut. It goes by many other names including English, French, Circassian, or Spanish walnut. It's called French walnut because that's where most of it used to come from, but most stockmakers are now using English walnut from northern California. This is a beautiful wood that characteristically has dark brown or black streaks running through a much lighter brown or even blonde background. Blanks currently range from $250 to $600 with some presentation marble cake (a cloudy smoky grain pattern) blanks going for much more. Good English walnut is very hard yet fairly light, and its tiny pores take precise inletting and checkering better than most woods.

The other walnuts are Claro, Bastogne and Black walnut. They have many good properties but are not as popular as English walnut with either stockmakers or customers. Black walnut is the least expensive; it costs about half that of an English blank. Claro is popular with customers wanting a showy piece of wood at reasonable cost. Bastogne walnut is a hybrid of English and Claro and has some properties of both. It is very strong and is good for heavy

Jere Eggleston does a bit of final inletting before shaping a stock of New Zealand walnut.

caliber rifles. Because it is a hybrid, it can cost as much as the better English blanks.

After your choice of an action and wood, stockmaker can begin his work. Let's see what happens from here, on a Mauser action for example, to understand what it takes to build a custom rifle.

1. Stockmaker lays out profile of the stock on the blank.
2. Action is set off to have barrel fitted.
3. Inletting is started.
4. Stockmaker machines or hand-works blank (steps 3 & 4 take 30 to 50 hours average).
5. Final sanding of wood in preparation for stock finish.
6. Stock finish is applied (if oil is used this can take up to 25 coats over a month).
7. Metalsmithing is done (usually by someone else, it may take from several months to a couple of years).
8. Checkering begins (30 to 50 hours average).
9. Gun is sent off to be engraved (may take months to more than a year).
10. Barreled action is blued (usually by someone other than stockmaker).
11. Action is received by stockmaker and put into finished stock.
12. Scope is fitted.
13. Rifle is sighted for accuracy and shipped.

All of this happens smoothly *only* if everything goes well for the stockmaker. Obviously a good deal of shipping time is involved. It is easy to see why it takes six months to a couple of years to get a finished gun. But then it has been done expressly for you and it's worth the wait.

Now that you know what it takes to have one of the fine guns built, who can you get to make one? There are perhaps two dozen rifle makers in the U.S. today. Historically most of these craftsmen have come from the Midwest and far West, however, I'd like to talk about three excellent craftsmen from the East Coast.

The first is Joe Balickie, a 48 year old full-time stockmaker working out of Apex, North Carolina. His work is admired nationwide and he is currently delivering in one to two years. Balickie is a former professional photographer who located in Apex 20 years ago. He grew up in a hunting family in the coal country of Hazelton, Pennsylvania. Although he has been a full-time maker for over 12 years, he has been working with guns for more than 30 years. About 12 guns a year go to customers and he builds three or four for speculation.

A tall, lean man, Balickie has the tanned face of someone who loves to be outdoors. Although an avid deer hunter, he is equally at home in the duck marsh. He spends an average of 100 hours per gun, and considers the crafting of a custom rifle an art form and a great hedge against inflation.

Like most rifle makers, he is somewhat opinionated about which actions to use. The Mauser and pre-'64 Model 70 are the only bolt actions he uses and he builds some guns on the Ruger No. 1 action.

"There has not been a good bolt action introduced since 1936 (Model 70)," he says pointedly. "When a gun company calls an action 'improved,' that really means easier to make."

Balickie, like many other stockmakers, uses a duplicating machine to do the basic shaping and inletting on his stocks. Although some purists don't like machine inletted stocks, Balickie doesn't agree. He believes it saves time and money for the customer, while leaving all of the hard work to be done by hand.

"The stocks are machined to my pattern so the customer is getting my gun. It just saves a little time and that makes sense," he points out.

Instead of a traditional oil finish, the North Carolina stockmaker uses a commercial varnish type finish that he feels has best moisture resistance. He also uses a thinned epoxy mixture on the inside wood for a good seal.

Balickie rifles are known for their light weight, slimness and thin grips. His lightweight Mausers average about 7 to 7½ lbs. and his Model 70 is from 8 to 8¼ lbs.

Checkering is his favorite part of riflemaking. He uses mostly 24 lines per inch. "It's the best compromise between aesthetics and using if the wood will take it," he says. "Smaller than 24 is more ornamental than useful." Except on request or on matched pairs, he always changes his checkering patterns. "Every one of my guns is different from the last one," he proudly adds.

Never at a loss for words on the subject of custom guns, Balickie has a special feeling for singleshots. He feels the Ruger No. 1 is the finest single shot ever designed from aesthetic and mechanical standpoints. "They're just flat out pretty," he says and adds with a grin, "I've never seen a pretty bolt action rifle, mine or anybody else's."

Although his guns are a good investment, going up at least 20 percent per year, Balickie says 95 percent of his customers hunt with their guns. As for price he is not taking orders at a quoted price but if a customer supplies all the materials, it will cost about $1,700 for a Balickie rifle. If the gunmaker supplies everything, it will cost from $3,200 up.

Jere Eggleston is a transplanted New Englander who builds fine rifles in his spare time in Columbia, South Carolina. He has been making rifle stocks since age 14 when he restocked a favorite .22. An avid hunter, he has taken ten species of North American big game including dall and stone sheep, in addition to two Scottish red deer.

Eggleston, although not well known nationally, is producing rifles on a par with anyone. He works on only one gun at a time, carrying it through more than 80 hours of work as he does everything by hand. He has very definite ideas on what constitutes a good rifle stock. For example, his stocks don't have a cheekpiece or fore-end tip. As for cheekpieces, he feels his high thick comb does the job just as well.

"A cheekpiece is pure ornamentation, a redundancy that can be left off. It doesn't make for a trim stock," he says.

Instead of a fore-end tip, Eggleston fashions a graceful schnabel. His love for schnabels goes back to boyhood years when he admired an uncle's fine German hunting rifles that had schnabels.

"I'll admit that my schnabel fore-end tip

Never at a loss for words on the subject of custom guns, Joe Balickie has a special feeling for single-shot guns. He feels the Ruger No. 1 is the finest single shot ever designed from aesthetic and mechanical standpoints. "They're just flat out pretty," he says and adds with a grin, "I've never seen a pretty bolt action rifle, mine or anybody else's."

may be considered a flourish by some," he says, "but to me it is a classic, graceful method of terminating the awkward end of a stock."

Special features on Eggleston rifles are a comfortably shaped grip and a skeleton grip cap which he makes from a 3/16 inch piece of steel. The inside of the grip cap is checkered.

Eggleston prefers the pre-'64 Model 70 or mauser actions. He is looking forward to crafting a rifle on the Ruger No. 1 action and thinks it may become a classic action of the future. He also feels very strongly about checkering and rubber butt pads.

"Who wants a rifle that won't stand up in the corner of a hunting cabin?" he asks. (Steel butt plates are prone to slip.)

Again looking to the practical side of a hunting rifle, Eggleston doesn't ordinarily use 24-to-26-lines-per-inch checkering. "I have found 20-lines-per-inch checkering as being the most practical and serviceable with 22 lines per inch almost as good."

With all the synthetic finishes so popular today, Eggleston still prefers the traditional oil finish. He puts up to 24 coats on the stock which takes more than a month to finish the job.

"The oil finish is the most tried and proven," the craftsman advises. "It holds up in the weather and is very abrasion resistant."

Eggleston charges $1,500 labor to craft one of his beautiful rifles if the customer supplies all of the components. Present delivery is about six to eight months.

Paul Jaeger, Inc. is rather unusual in the custom rifle business. The firm under the direction of Dietrich Apel, who is a fourth generation gunsmith and a nephew of Paul Jaeger who started the company over 50 years ago. What makes Jaeger so unique is that all of the gunsmithing connected to a fine custom rifle is handled under one roof. There are gunsmiths from America, Germany, England, Switzerland and Austria working at Jaeger's Jenkinstown, Pennsylvania shop. They are busy crafting what Apel calls the "American modern classic" hunting rifle. They can produce within limits just about anything a customer might want in a classic stocked rifle. Skeleton butt plates, steel butt plates, rubber pads, ebony fore-end tip, schnabel fore-ends, plain fore-ends, all types of checkering, etc. are yours for the asking. They use model 70, FN, Mauser, Ruger No. 1, Remington 700 lefthand, and other actions according to the taste of the customer.

Under Apel, Swiss-born Alfred Wyss-Gallifent runs the custom rifle making at Jaeger. Engraver Claus Willig and checkerer Dennis Richards work at home for Jaeger.

Dietrich Apel, with a slight Old-Country accent, says the bolt action rifles produced today are the best ever. He thinks his rifles are a good investment in many ways. "They give the owner a lifetime of pleasure and are something he can give to his children."

Jaeger craftsmen are now producing what they call their Signature Grade rifle. Limited to 50 pieces, it is simply their best effort with any action the customer wants. The rifles will run in price from $2,500 without engraving to over $10,000 for a fully engraved gun. Other custom rifles from Jaeger average $3,000 to $4,000. They have the traditional oil finish and 24-lines-per-inch checkering. Almost all of the guns are stocked with French walnut and delivery is from six months to a year.

Whether ordered from these fine gunmakers or from one of the other top craftsmen around the country, the American classic stocked rifle has evolved into truly one of the world's great guns. ♦

DOUBLE Visions

Double-barreled rifles conjure images from another age ... when intrepid sportsmen traveled to the far corners of the globe to pursue the world's most dangerous game.

MICHAEL McINTOSH

These guns conjure visions. Of Cape buffalo with wicked, double-curved horns. Of acacia trees and thornbrush, and Kilimanjaro's cloudy crown. Of the Indian forest, fresh pug marks, a creaky *machan*, and great, striped cats in the dappled shade.

Among all the products of the gun-maker's art, none display more patient craftsmanship, more consummate skill, or a greater aura of sheer romance than a double rifle. The earliest were built in Europe, particularly in Germany, Austria, and Hungary. Forest game, from deer and hare to bear and wild boar, traditionally was hunted by driving. This called for a gun that would be fairly short, highly dynamic, as carefully fitted as a fine shotgun, and capable of an instant second shot. A double was the obvious answer, both for handling qualities and firepower, since it would be 200 years or more before any magazine rifle could handle cartridges large enough for the world's biggest game.

Even though Continental makers continued to turn out quality specimens, the double rifle reached the peak of its refinement in England. As they did with the double shotgun, the British gun-makers took an already-existing form and perfected it to an extent that a top-quality double rifle is, by definition, an English specialty.

Some were built as flintlocks, but double rifles did not come into wide use as sporting arms until the 1840s, in the early years of the percussion system. The majority of these muzzle-loading guns amounted to little more than shotguns fitted with rifle sights; they were smooth-bored with bore measurements expressed in gauge rather than caliber. The most common were 12, 14, and 16 bores, all meant for use with round lead balls and three to five drams of black powder. They were accurate only to about 60 yards and, despite the size of the slugs they fired, not particularly powerful.

With the coming of the great arms-making revolution of the 1850s, the breech-loading double rifle evolved along much the same lines as the sporting shotgun. By 1860, virtually every gun-maker in England was experimenting with some sort of break-action shotgun. Similar advances in self-contained ammunition opened a whole new world of possibilities for improving the rifle. Improvements in cartridges came more slowly though, and the typical double rifle of the early breech-loading days still had smooth bores and fired shotgun-sized slugs from brass cases. But the 1860s and 1870s brought almost daily advances in design and ballistics, and the modern rifle began taking shape long before its predecessor became truly obsolete. The experiences of English sportsmen in far-flung corners of the world helped the pace along.

With the expansion of the British Empire, hunters in Africa, India, and the Far East found themselves face to face — and often enough, nose to nose — with game that got mightily annoyed when shot with smooth-bored, pumpkin-ball guns. The stories of most of the chaps who went out with 12-bore guns looking for elephant or rhinoceros, Cape buffalo, or gaur, were written years later and in the past tense. These men were admired more for their enthusiasm than for their success.

At first, sportsmen bound for the colonies, especially those not keen on being stomped into marmalade, simply ordered bigger guns — 10, 8, and even 4 bores, of both muzzle- and breech-loading types. But about the same time, some makers began experiments along another path, one which eventually led to the modern rifle.

In 1856, James Purdey introduced his "Express Train" rifles — muzzle-loaders with two rifling grooves in each barrel. They were small-bores by contemporary standards — calibers .450, .500, and .577 — but Purdey's notion was that smaller bullets at higher velocities were the key. As other makers took up the idea and applied it to cartridge rifles, the name was shortened to "Express" and eventually came to signify the whole class of rifles in the mid-caliber range of .400 to .600.

Unfortunately, big game animals were no more impressed by fanciful names than they were by small, fast-moving bullets. And so, the big-bore rifle — a massive piece of work with exposed hammers and short, 8-bore barrels — remained an essential item in the African and Asian arsenal. But the invention of Cordite in 1885 ultimately proved Purdey right. Cordite was considerably more powerful than black powder and provided the fuel to give relatively small bullets enough punch to put a big animal down for keeps. From then until the beginning of World War II, the English double rifle flourished in its golden age.

Dozens of gun-makers have built double rifles and all have offered them strictly as made-to-order items incorporating virtually any feature the customer wanted. As a result, double rifles cover an astonishing variety of actions, bolting systems, rifling techniques, cartridges, and calibers. Unless they were specifically ordered as a matched pair, no two are alike.

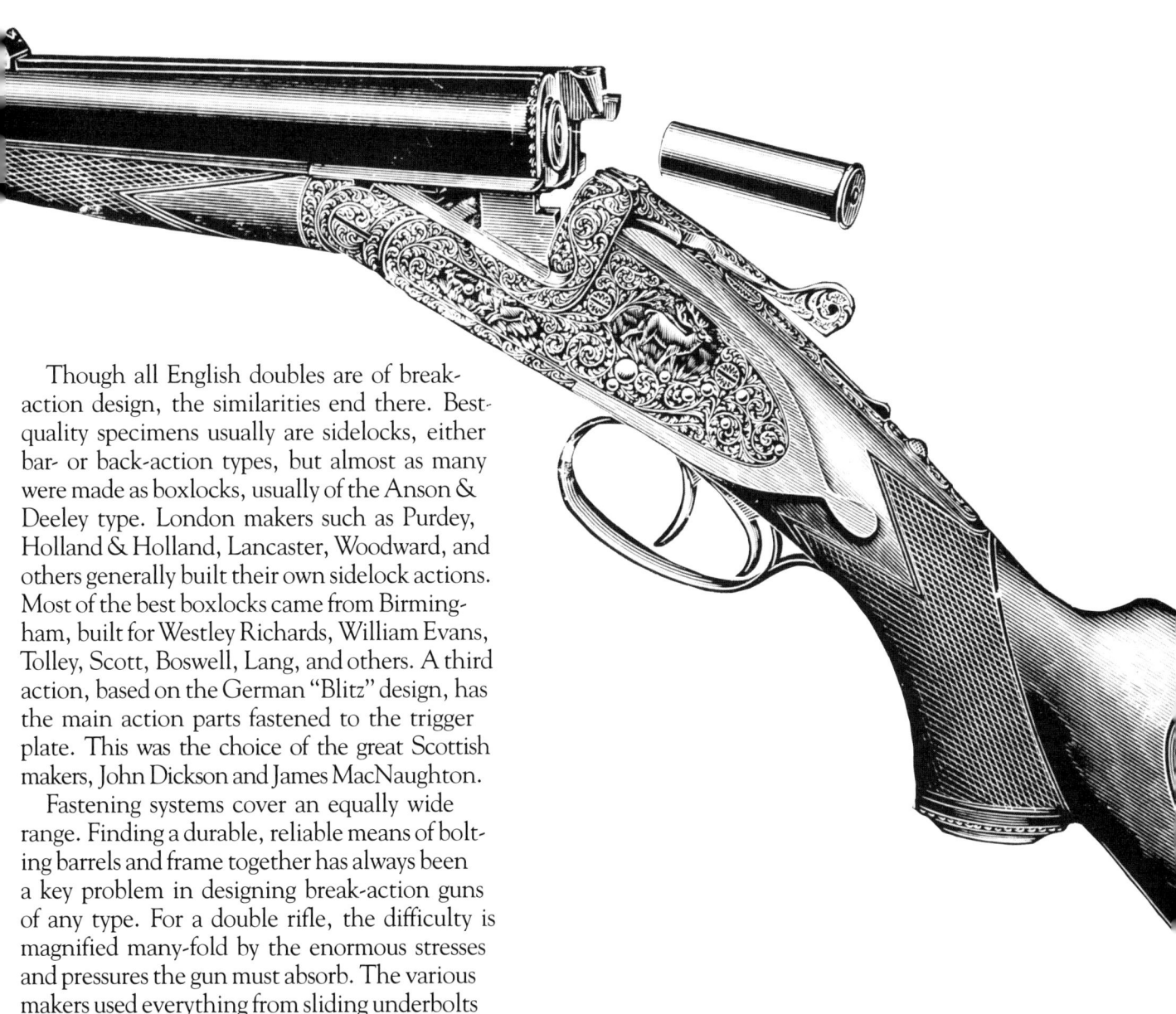

Though all English doubles are of break-action design, the similarities end there. Best-quality specimens usually are sidelocks, either bar- or back-action types, but almost as many were made as boxlocks, usually of the Anson & Deeley type. London makers such as Purdey, Holland & Holland, Lancaster, Woodward, and others generally built their own sidelock actions. Most of the best boxlocks came from Birmingham, built for Westley Richards, William Evans, Tolley, Scott, Boswell, Lang, and others. A third action, based on the German "Blitz" design, has the main action parts fastened to the trigger plate. This was the choice of the great Scottish makers, John Dickson and James MacNaughton.

Fastening systems cover an equally wide range. Finding a durable, reliable means of bolting barrels and frame together has always been a key problem in designing break-action guns of any type. For a double rifle, the difficulty is magnified many-fold by the enormous stresses and pressures the gun must absorb. The various makers used everything from sliding underbolts and doll's-head extensions, to cams and single

Drawing from a 1922 Webley & Scott Ltd. catalogue.

The invention of Cordite, which propels a bullet at a much higher speed than black powder, heralded the golden age of English double rifles. Cordite also paved the way for a seemingly endless array of cartridges, everything from .22 Hornet to .45-70 and beyond.

and double cross-bolts, both round and square. Continental makers, especially the Germans and Austrians, have always been fond of the cross-bolt, usually the double, Kersten-type, and at least one English maker — W. W. Greener — used it, too. The Greener cross-bolt is a single round pin mounted horizontally in the frame at a right angle to the barrels, and it engages a hole in a massive rib extension. Time has proven it a successful design both for shotguns and rifles.

The most common fastening systems are the triple-bite arrangements: a double underlug and a third lug bearing on a rib extension; and the Purdey-type, which also uses a twin underbolt in combination with a small lug between the barrels that fits into a recess in the standing breech.

Some double rifles, like some shotguns, are self-openers. They are fitted with two spring-loaded pins in the water table that help start the barrels swinging down on the hinge pin when the fastening bolts are retracted. The only practical purpose of this in a shotgun is to make loading a bit quicker. But it has a real advantage in a rifle. Chamber pressure generated by a huge cartridge tends to force the case back against the standing breech so hard that a gun can be extremely hard to open. Any hunter who ever had to reload his double with a big animal about to dance on his head no doubt appreciated any mechanical advantage the maker could build in.

The bolting mechanism of most early doubles was operated by a lever located under the frame. The lever usually curved around the trigger guard, keeping it as close as possible to the shooter's hand. It swung sideways to release the bolts. Later, the mid- and small-bore rifles were more commonly fitted with top latches of the conventional shotgun type, but the really big guns, 8-bores and the ilk, generally were made with underlevers and exposed hammers long after the hammerless, top-latch style became predominant. The top latch is faster to operate, but underlevers had an advantage in that they could be operated in complete silence. Some, notably the Woodwards, were designed so the underlever was pushed forward rather than swung to the side. This exerted some camming pressure against the barrels, and helped to overcome sticky cases — and probably a few sticky wickets besides.

Long before the English double evolved to anything even near its perfected form, armsmakers the world over knew that a spinning bullet flies truer than one that's simply lobbed into space. Like choke-boring in shotgun barrels, the theory and practice of rifling was no doubt discovered, forgotten, and rediscovered many times in many places before it finally came into universal use. Predictably, the bores of double rifles have been treated to every technique from smooth-boring to full-length rifling and to any number of curious variations in between. The most successful ones cover the whole range.

Smooth-bores came first and within their limitations, they performed surprisingly well. They fired monstrous slugs that fit the bores loosely and were meant for very close-range work. Their heyday, however, was brief, and they were soon made obsolete by a bewildering array of rifling methods.

Every maker seems to have had his pet theory about how to make a bullet spin. Variations in the number, and the width and the depth of grooves and lands are endless. Some makers bored two deep grooves opposite each other and used bullets with flanges that rode the grooves like splines in a gear. Others cut many square-bottomed grooves, and yet others rounded the grooves so the lands were pointed rather than flat. Some grooves were shallow, others deep.

Rigby popularized a form called ratchet rifling, with each groove cut deeper on one side than on the other. In cross section, the bore has the shape of a ratchet gear. The method worked quite well, especially if the deeper parts of the grooves were opposite the twist of the rifling, which gave the bullet a good spin without much friction against the bore.

Possibly the weirdest rifling was Charles Lancaster's oval bore, an old notion that Lancaster took up and eventually perfected. Instead of a straight bore with spiral grooves cut into it, the oval bore was smooth and slightly egg-shaped, and the bore itself was cut in a shallow spiral, a sort of auger-bit principle in reverse. Lancaster bored these barrels with a progressively faster rate of twist from breech to muzzle, and the whole thing performed surprisingly well.

One of the most popular treatments around the turn of the century was Colonel George Vincent Fosbery's patented combination of smoothbore and rifling. Holland & Holland gave it its most common name — Paradox. And paradox it was: a barrel bored perfectly smooth through most of its length, choked like a shotgun, and grooved with two to four inches of shallow rifling at the muzzle. Such barrels handled both ball and shot so successfully that virtually every maker in England eventually offered at least one model of gun with muzzle rifling.

Except for the very earliest ones, double rifles never were strictly big-bore items, and the coming of the Cordite age opened the way to an endless variety of calibers. Many of the cartridges originated from some hunter's notion of what a proper load ought to be; such ammunition was custom-made along with the gun to fire it. Others were proprietary, developed and standardized by one maker or another. Between 1897 and 1939, Holland & Holland, John Rigby, Purdey, Westley Richards, Gibbs, and W. J. Jeffery introduced more than 25 new cartridges, some of which became standard fare.

But given rifles made to order and craftsmen of almost sublime art, there were few limits. The bottom line simply was that you could have almost anything you could afford. If none of the standard cartridges suited your fancy, you could invent your own. Lots of double rifles were built for Americans and chambered for American cartridges, usually those made up with rimmed cases. Search long enough and you'll turn up fine English doubles chambered for everything from .22 Hornet to .45-70 and beyond. Continental makers developed almost as many different cartridges as the British.

The whole point of a double rifle is to have two guns in one: two barrels, two locks, two triggers, two ejectors. These rifles were developed at a time when sportsmen traveled literally to the ends of the earth to pursue the world's most dangerous game. They needed dynamic handling and utter reliability, and makers took great pains to provide them. Some offered their best-quality rifles with extra sets of locks that could be installed without tools. The Holland & Holland hand-detachable sidelocks are probably the most famous. In the best Westley Richards boxlocks, the trigger plate can be unlatched, removed, and new locks slid in by hand in a matter of seconds. Daniel Fraser of Edinburgh, whose rifles are few but of superb quality, plated all internal action and ejector parts with gold for smooth operation and for protection against rust in tropical climates.

Few double rifles were made with single triggers. Most makers simply refused to build them, arguing that a one-triggered gun could be rendered utterly useless by a malfunction. Westley Richards was the first and for a long time the only important maker to offer a single trigger, and it apparently was a good one. James Sutherland, the first Englishman to kill 1,000 elephants, used a pair of Westley Richards .577 doubles, each with only one trigger.

The problems endemic to building a double rifle are similar in kind, but quite different in degree from those of a double shotgun. Obviously, the means of fastening the action must be strong and certain. But the real trick is in making two barrels, mounted side by side, that shoot to the same mark at a given distance. Collimating shotgun barrels is difficult enough, even though a spreading shot swarm permits some latitude. Not so with a rifle. Close isn't good enough, and the most critical phase of building a double is what the British call shooting-in.

Like those of a shotgun, double rifle barrels are laid in at a slight angle rather than parallel, so that bore centers converge at a given distance. Beyond that distance, doubles naturally cross-fire. Even though many are fitted with a series of folding leaf sights that may be staged

The trick in making a double rifle is to position the barrels so they shoot to the same mark. By the early 1960s, only four men in all of England excelled at the art of shooting-in a double rifle.

to as much as 500 yards, every double rifle's accuracy deteriorates at a geometric rate as the bullets exceed the distance for which the barrels were regulated.

First step in the regulating process is to braze the barrels together at the breech end and soft-solder them at the muzzles. They are then fitted to the stocked frame and taken to a shooting range. Customarily working with a 100-yard mark, the shooter-in fires shot after shot, carefully adjusting each barrel's point of impact by driving slender wedges between the muzzles or by closing the gap between them. Once shot in, the barrels are brazed together full-length and finished.

Shooting-in is more art than craft. By the early 1960s, only four men in England were capable of the job and two of them worked at the Holland & Holland Shooting School outside London. There are probably even fewer qualified shooters-in today.

Which has been the fate of most of the skills necessary for building fine double guns, particularly rifles. Once you could request a custom-made double rifle from any of more than 30 English makers. Now there are scarcely more than two: Purdey and Holland & Holland. W&C Scott discontinued double rifles in 1940. John Wilkes, one of the last independent London makers, stopped making them years ago. You can still have doubles built on the continent — rifles of a basic quality that will stand well in any company. But they aren't London's best.

There is no snobbery. The simple truth is that no gunmakers in the world have ever combined utility, quality, and aesthetics to quite the extent that the English have. But the time of the British double rifle is past. Magazine rifles now can accommodate more powerful cartridges at a fraction of the manufacturing cost. A fine English double is an artifact and the wonderful skill required to build one probably will not exist in another generation or two.

A fine double is also a piece of magic. Hold one in your hands and you'll see the visions: of elephants with a hundred pounds of ivory to a side; of a lion's baleful eyes staring through sun-bleached grass; of a nine-foot tiger in an ancient jungle. Of a time we never knew and a time we'll never see again. ♦

The Surprising 28
Sometimes A Great Notion

Besides being eminently collectible, the old 28 bores can be great fun to shoot — light enough to carry all day, but heavy enough to handle with control.

MICHAEL McINTOSH

It was clear by 1880 that small-bore breechloaders could be effective game guns. The London Gun Trials of 1875 demonstrated that a well-made 20 bore in the hands of a competent shot could more than hold its own against larger guns inside 40 yards, and subsequent experience proved that even smaller gauges could deliver some surprising results. Greener, with a mixture of pride and bemusement, reported that an 11-year-old set the London shooting circles abuzz by killing as many as 38 of 50 best Blue Rock pigeons at 27 yards rise with a 28 gauge. Greener's pride came from the fact that he'd built the gun, his bemusement from the almost universal belief that really small-gauge guns aren't supposed to shoot that well. Another London gunner reported similar results at game, writing to Greener in 1885: "I can only say your 28-bore gun cannot be improved upon."

Many shooters who had never tried a 28 gauge scoffed. Many still do. And the 28 gauge still has its surprises.

It was a relative latecomer to America and caused no great stir even when it did arrive. In fact, the 28 gauge had so little impact upon American gunning that most of the great makers

---

Elegant 28s on today's market include from top: Parker Reproduction D Grade, 26-inch barrel with splinter forend and English stock; Beretta S687EELL, 26-inch barrel, hand-engraved with European walnut stock. Photograph by Art Carter

either ignored it altogether or waited until skeet shooting created a market, almost a generation after the first American 28s were built. Neither A.H. Fox, L.C. Smith, Lefever nor Baker ever put a 28 gauge into production.

Parker made the first American 28-gauge doubles about 1905. Around the same time, Remington chambered its Rider single-barrel in 28 gauge, but the greater share of the small-bore market, which in those days was every bit as brisk as a tombstone, went to the double.

Parker's first serious competition came from Ithaca, who offered the 28 gauge in 1911 as a chambering for its Flues Model double. When the New Ithaca Double was put into production in 1925, it, too, was made as a 28.

Both the Parker and Ithaca are true 28 bores, built on frames scaled to suit the slender barrels. Parker built most of its 28s on No. 00 frames but also used the No. 0 frame, standard for 20 gauge, in order to get proper balance in 28s ordered with particularly long barrels, which could be 24, 26, 28, or 30 inches. All Parkers tend to be relatively heavy, and the 28 gauges generally weigh from 5¾ to 6 pounds. Flues Model Ithacas are the lightest of all American 28-gauge doubles — from 4¾ to 5¼ pounds, with either 24- or 26-inch barrels. New Ithaca Doubles are a bit heavier and could have 26- or 28-inch barrels, but at 5¾ pounds, a 28-gauge NID won't wear you down if you carry it all day.

Harrington & Richardson built some single-barrel hammer guns in 28 gauge during the 1910s and '20s, but the rest of the arms industry made no effort to adopt the little shell until the 1930s, when skeet shooting blossomed. There, the 28 finally found a home.

By then, the double gun was in deep doldrums, and most of the 28 gauges built in America over the next 50 years would be repeaters. Winchester brought out a 28-gauge version of the Model 12 in November 1934, probably the first such repeater built anywhere in the world and certainly the first made to 28-gauge scale. The earliest guns were available with 26- or 28-inch barrels, either plain or solid-ribbed, and weighed about 6¼ pounds. Ventilated ribs came along shortly after. The Model 12 ultimately was made in 15 different styles of 28-gauge field and skeet guns, a dozen of which were discontinued in 1945, two more in 1955, and the last in 1959.

In 1936, Winchester offered the Model 21 in 28 gauge as a special-order item. Like the .410s that came along nearly two decades later, the 28-gauge Model 21s were built on 20-gauge frames. Iver Johnson chambered its Skeeter side-

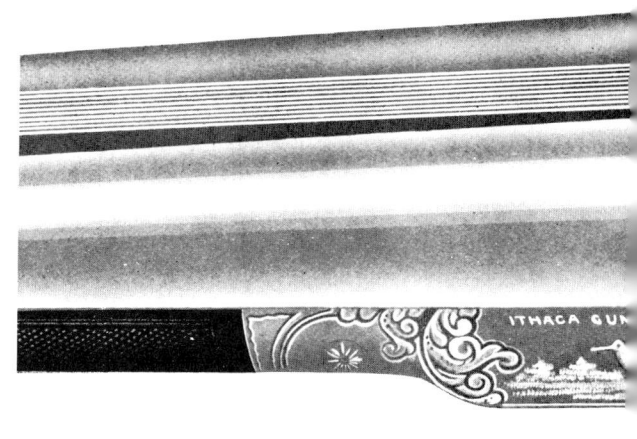

Page from the 1922 Ithaca catalogue features the "neat and tasty" No. 3 double, produced in 28 gauge with a 26-inch barrel. Note the $80 price, which included War Tax.

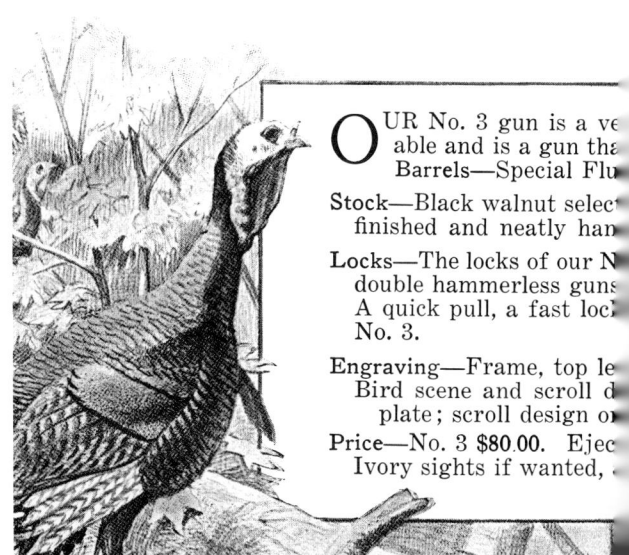

by-side in 28 gauge during the '30s. If you had the money, you could get a European-built 28 — a Beretta, a Francotte, a German-made Charles Daley, a Merkel, or a gun from any of the best English makers. Still, the 28's niche in the American market remained a tiny one.

Shooting sports of all kinds enjoyed something of a renaissance following World War II, and a new vitality rippled through the arms-making industry. Even though the general trend favored big-bore guns and lots of firepower, the 28 gauge got a share of attention as both a skeet and game gun. In 1952 Remington brought out the first 28-gauge autoloader, a scaled-down Model 11-48. Parker and Ithaca doubles were out of production for good, and if you wanted a new 28 double, it had to be either a Model

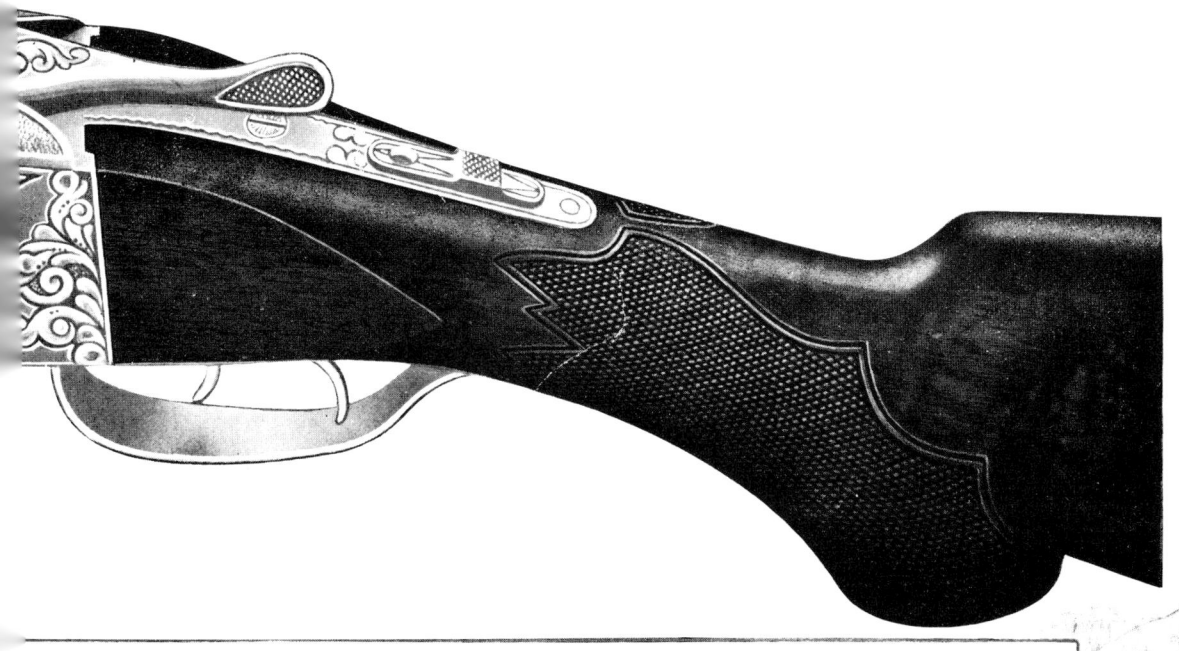

y gun. It is a medium-priced gun, well made, durable and dependmay be proud of in any company.

s made in full pistol grip unless otherwise ordered, it is well fitted, the stock is let into frame to prevent splitting and spreading.

e same mechanical construction and material as our higher grade

working and simplicity of construction are characteristics of the

trigger plate and guard engraved by hand.

nd left sides of frame; duck scene and scroll design on trigger nd scroll design on guard.

ts $15.00 extra. Soft rubber recoil pad if wanted, $5.50. Lyman

Specifications

Full pistol grip.

28 gauge, 26 in. barrels.
20 gauge, 26 or 28 in. barrels.
16 gauge, 26, 28 or 30 in. barrels.
12 gauge, 26, 28, 30 or 32 in. barrels.
10 gauge, 32 in. barrels.
Weight—10 ga., 9½ to 10 lbs.; 12 ga., 6¾ to 8¼ lbs.; 16 ga., 6 to 6¾ lbs.; 20 ga., 5¾ to 6¼ lbs.; 28 ga., 5½ to 5¾ lbs.

21 Winchester or be imported from Europe.

In the 1960s the 28 bore fared better than it ever had, at least in the variety of available guns. Browning introduced the 28-gauge Superposed in 1960. Before the decade was over, 28s were available as over/unders from Charles Daley (by then built in Japan), Krieghoff, SKB-Ithaca, and Winchester. None were built on true 28-gauge frames. You could get side-by-sides from Webley & Scott, Beretta, Darne, Franchi, and Galef, inexpensive single-barrels from Winchester and Stevens. High Standard introduced a 28-gauge pump in 1966. Remington adapted both its Model 870 pump gun and its Model 1100 autoloader to 28 gauge in 1969.

Today, the picture is brighter yet. The Remington 1100 and 870 are the only 28-gauge guns currently made in America, but both Winchester and Browning offer Japanese-built over/unders. Winchester's Model 23 side-by-side, also made in Japan, is now available in a classic series as two-barrel sets — 12 and 20, 28 and .410. The remarkable Parker Reproduction, yet another Japanese item, is the old Parker brought back to life. The 28 gauge, on the market now for a year, is built on the old standard No. 00 frame. On the import market, you can get a 28 from any of almost two dozen European markers. Most are side-by-sides. Beretta makes a production over/under in 28 gauge, and you can bespeak a custom over/under from the best English and Italian makers if you're feeling flush. Costs of the imports range from cheap to serious, and so does quality.

Still, not all that many 28-gauge guns are actually built. The skeet world, which comprises the majority of 28-gauge shooters, has largely abandoned the notion of separate guns for each events. Most serious skeeters nowadays use either multi-barrel sets — 20, 28, and .410 barrels on 12-gauge frames — or interchangeable barrel tubes for the small gauges. To shoot the small-bore events, you simply slide the proper set of tubes in your 12-gauge gun and have at it. Tubes certainly are an economical alternative: a full set of small-gauge tubes from Briley or Kolar for $1,000 or less versus four guns from Perazzi or Marocchi or Shotguns of Ulm at about $2,000 apiece. And you can't buy them anyway, because none of the best-quality target guns are available as small-bores built to scale. So, if you want to compete seriously in all skeet events, you shoot small-bores that weigh more than 12 gauges. Tubes are a poor compromise for anyone who takes delight in the handling qualities of the various gauges, but they work.

Twenty-eight gauge specimens of the great

If you don't want a made-in-America Remington or Japanese-built Winchester or Browning, you can get a 28 from any of almost two dozen European makers. Costs of the imports range from cheap to serious, and so does quality.

American doubles are collector's darlings. No maker built very many of them. The rarest of all is L.C. Smith — or rather *the* L.C. Smith, because there was only one, an 00 Grade Hunter Arms gun, serial number 100. It probably was built around 1910. At last report, the Hunter family still owned it.

Factory records no longer exist, but Winchester estimates that no more than 200 Model 21 production guns were built as 28 bores. Even on 20-gauge frames, they're nifty little things, and hindsight still nettles me for turning down the one I could have bought for $2,000 some 15 years ago.

Ithaca built more 28s than any other American maker. No record remains of how many Flues Models there were, but records do exist for the New Ithaca Double series. Between 1925 and 1948, when the NIDs went out of production, Ithaca made 415: 295 of them Field Grade, 42 in No. 1, 58 in No. 2, five No. 3s, ten No. 4s, three in No. 5, and two No. 7s. Curiously, there are more than twice as many .410 NIDs than 28 gauges.

Not surprisingly, Parkers are the most sought-after and fetch the highest prices. Only Magnum 10-gauge and .410 Parkers are rarer. The most famous of the Parker 28s is the A-1 Special that somehow came to be called Little Persuader (which gets my vote as the dumbest name ever given a shotgun); it's the gun that brought $95,000 at a Christie's auction in 1981. It has sold two or three times since then, and you can bet the price hasn't gone down. Less well known is the DHE 28 gauge that Carole Lombard ordered in 1936 as a gift for Clark Gable. It's still in virtually mint condition and, except for the initials CG, is a typical Remington Parker — but those initials lend it a human connection more interesting than any price tag.

Besides being eminently collectible, the old 28 bores can be great fun to shoot. But choose wisely if you decide to do that, for several reasons. In any reasonable condition, a 28-gauge Parker or Ithaca is likely to bring a premium price, and though further use of a gun that's already been well used won't necessarily lower the value, it won't raise it, either. The question of use is best answered gun by gun in terms of originality and condition.

The 28 finally found a home in the 1930s, when skeet shooting blossomed in America. Up to that time most 28-gauge guns were doubles, but for the next half-century most would be repeaters.

Age, too, is a consideration, because in the old days the standard 28-gauge shell was 2½ inches long, and the earliest guns were so chambered. The original load was 1¾ drams of black powder (or its equivalent in smokeless) and ⅝-ounce of shot. Sometime after World War I, ammunition makers developed a 2⅞-inch cartridge that held a bit more powder and ¾-ounce of shot. The now-standard 2¾-inch shells came along in the 1930s as skeet loads; they also held ¾-ounce of shot. All three cartridges were readily available until World War II.

Ithaca began boring all its 28-gauge chambers to 2⅞ inches in September 1931. Parker did the same thing, but I don't know exactly when. All 28-gauge Model 12s have 2⅞-inch chambers, but some of the Model 21s built before World War II are bored 2½ inches. It's wise to have any older 28 miked out by a good gunsmith.

If you've never shot a 28 gauge, you're in for a treat, provided you don't ask more of it than it's capable of doing. Unfortunately, the 28 is often thought of as similar to the .410; in fact, they are vastly different, and the 28 delivers far better, far more consistent performance. The standard 28-gauge bore, nominally .550 inches in diameter, and the standard ¾-ounce shot charge seem to suit one another beautifully. The shot column is short enough to promote relatively little stringing, especially with hard shot, and it contains enough pellets (about 440 No. 9s and about 260 No. 7½s) to deliver efficient pattern density out to about 35 yards. Much beyond that, patterns get patchy, because a 28 just doesn't hold enough shot for long-range work. Years ago, Federal Cartridge made 1-ounce and, more recently, ⅞-ounce loads in 28 gauge, but neither made much sense. Increasing a shot charge without also increasing bore diameter is no improvement. Twenty-gauge loads belong in 20-bore guns.

Those who like to effuse upon its virtues sometimes are loath to admit that the 28 gauge has survived for reasons other than its intrinsic merit. Merit it has, but the fact remains that if there were no 28-gauge event in skeet, there would be no 28 gauge. Witness the demise of the 16. Intrinsic merit alone is not enough. A 28 is a wonderful target gun and, within its limitations, an equally good game gun, but the 20 gauge is more versatile and current economic realities are such that no American maker can produce a true 28 gauge — by which I mean a perfectly proportioned double — at reasonable cost. As a separately manufactured item, the 28 simply cannot sell enough copies to support itself. Consequently, most of the readily accessible 28s are 20-gauge guns with 28-gauge barrels.

Which is not as lamentable as it might seem. A large measure of the 28's appeal is that it can be lightweight and still comfortable to shoot. The rule of thumb among English makers is that a gun should be at least 96 times heavier than the shot charge it fires. For the 28, that means a gun of at least 4½ pounds, which is enough to dampen recoil but not necessarily enough to point and swing consistently. I've tried shotguns that light, but I'd hate to have to feed myself on the game I could bag with one. A 5½-pound gun, though, is another story, light enough to carry all day and still heavy enough to handle with control. And that's where the 28 gauge truly shines. ♦

Commemorative WINCHESTERS

Since their introduction in 1964, Winchester's commemorative rifles have become "instant collectibles." Now numbering over 878,000, the popular rifles have established a rich tradition and history of their own.

RICK HACKER

It seems like a classic test for a college marketing class: how do you take a best-selling product that already has an above average reputation and then make it even more desirable? In the case of Winchester Repeating Arms Company, the answer is now a part of firearms history — you make it a commemorative.

Winchester was not the first to come up with the commemorative concept; that honor must go to Colt. However, the historic repeating rifle company simply became the most prolific, due to an aggressive and imaginative marketing department that has turned out commemorative rifles every year since 1966. To date, over 878,000 rifles representing more than 40 individuals or events have been created and, under the new ownership of the U.S. Repeating Arms Company, the company's commemorative rifle program looks like it will continue into the decade of the 80s.

Actually, the idea for the first Winchester commemorative limited edition rifle did not come from the company at all, but was the product of the Wyoming Diamond Jubilee Commission, which commissioned 1,500 special order Model 94 carbines from the company in 1964. That was the first year of Winchester's somewhat unpopular decision to change manufacturing procedures for some of their rifles, the venerable 94 included. Ironically, although standard "post-64" Model 94s do not command high prices on the used gun market, the 1964

Wyoming Diamond Jubilee is one of the rarest commemoratives of all and commands a premium, although in light of recent developments, is certainly not the most expensive commemorative. Still, it is unique because it was the first of the limited edition Winchesters.

Mechanically, the Wyoming Jubilee was a standard Winchester 1964-era carbine, but with some noticeable external differences. For one, it was the first 94 carbine to be made with a saddle ring since the 1930s. Second, it sported a color casehardened receiver which was roll engraved with a simplified curl design. A specially-struck gold plated medallion was inset in the stock and the loading gate was also plated. In addition to its low production run, the fact that the gun was only available through one Wyoming sporting goods store added to its desirability. The Wyoming Diamond Jubilee carbine carried a list price of $99.95, but it's a substantiated fact that many of the guns sold for more than that to anxious collectors. Because the gun was an instant sellout and sold for approximately one-third more than the standard Model 94 carbine, Winchester's management gave the incident some serious thought.

They realized that Winchester's 100th anniversary was only two years away. For a firearms company whose products had helped tame the West, the magic years 1866-1966 were too good of an opportunity to pass up. Why not create a commemorative Winchester, such as

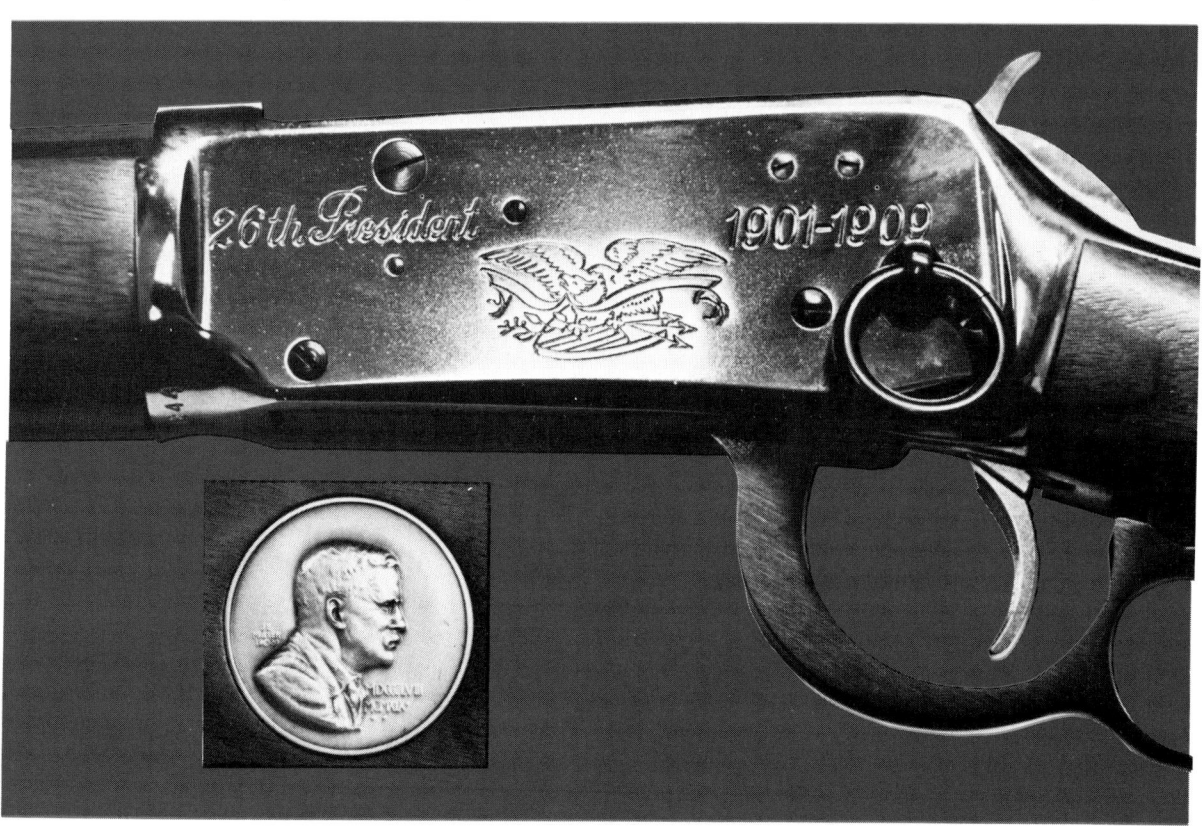

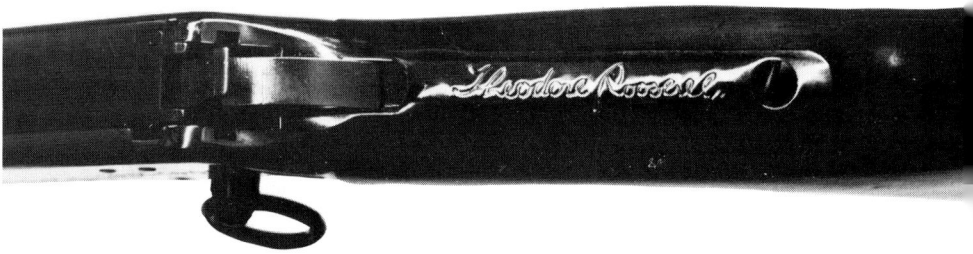

The Theodore Roosevelt commemorative is unique in that it bears the president's signature along the tang.

the Wyoming Jubilee folks did — to pay tribute to a historical past? Naturally, it should be a lever action, the style of gun most synonymous with the Winchester name. Unfortunately, the original Model 1866 brass framed "yellow boy" and its tooling were things of the past. However, the Model 94 was still in the line and at that time was well past the two million mark, making it the longest produced lever gun in the company's history. Using the Wyoming Jubilee commemorative as a precedent, the Model 94 thus became the basic action for all Winchester commemoratives produced from that day on.

Therefore, in January 1966, the Winchester Centennial Model '66 Carbine and Rifle commemoratives were announced to a surprised and unsuspecting market. Most shooters did not know what to expect of the unique lever-action. Should it be shot, or merely hung on the wall? The $125 price tag dictated that it was something special, yet the fact that over 100,000 of the Centennials were being produced seemed to say that the rifle was not as unique as Winchester's four-color ads hinted. Still, firearms aficionados bought the gun, although not as fervently as Winchester's marketing department may have wished. Even through 1967, unclaimed Centennials could still be found on dealers' shelves, but eventually, they all sold out. "Post-64" discrepancies notwithstanding, the Model '66 Centennial was an attractive gun. Both rifle and carbine featured a gold plated receiver and fore end cap, brass crescent buttplate, and blued octagon barrel, into which was roll-stamped. "A Century of Leadership 1866-1966."

During the same year the '66 Centennial made its appearance, another Winchester 94 commemorative was produced, but with far less fanfare. Known as the Nebraska Centennial Carbine, it was a standard Model 94, but accented with a gold plated forearm band and loading gate. The receiver bore the gold filled, roll-stamped legend, "Nebraska Centennial 1867-1967" on the lefthand side. Since only 2,500 of these carbines were produced, they quickly sold out and have since become the object of many a collector's search, due to their early appearance in Winchester's commemorative history, plus their relatively low production run.

Thus, in these two initial Winchester commemoratives, the basic rules of limited edition collecting was established! The lower the production run, the more rapidly the gun would sell. Yet, commemoratives produced in greater quantities could be obtained by a larger audience, thereby whetting the appetites of these collectors to seek out additional commemoratives as they became available. Initial retail price seems to have little effect on the collectibility of any Winchester commemorative, as almost all will eventually be sold within a given period of time. The Buffalo Bill Commemorative, of which 112,923 units were produced, had the highest production run of any "limited edition" Winchester, yet even they eventually sold out within three years at $130 each. However, the 1976 issue of the U.S. Bicentennial Winchester, of which only 19,999 were made, was quickly snatched up within a year at $325 per rifle.

Clearly, one of the real lures of the Winchester commemorative lies in the total number of units being produced; scarcity has always been a factor in collectibility. Yet there is another element which defies all logic and perhaps reveals the very heart of the firearms collector. That one intangible attraction is subject matter. The Illinois Susquicentennial,

---

The receiver of the Theodore Roosevelt commemorative is plated with white gold. Commemoratives are truly works of art, as exemplified by this special die-struck medallion, taken from a design by J.E. Fraser in 1920, a year after Roosevelt's death.

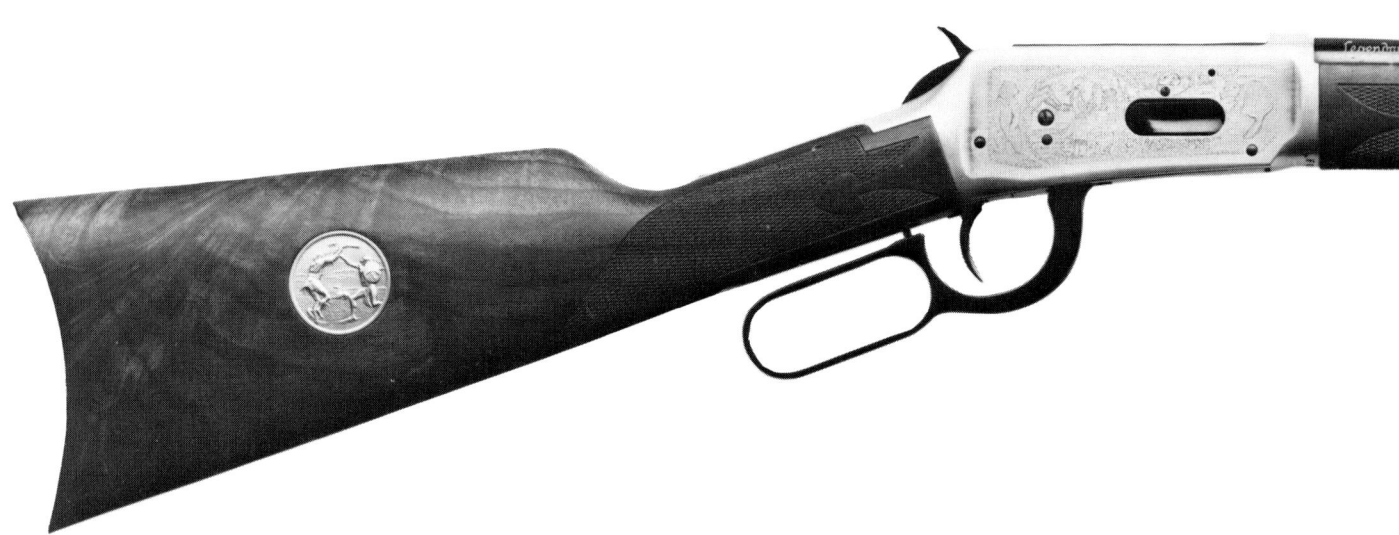

of which slightly over 37,000 were produced, appealed to a very limited number of collectors, even though this gun was sold and advertised nationally. Yet when 50,000 John Wayne Commemoratives (JWC) were offered to an eagerly waiting nation full of Duke's admirers in January of 1982, the gun quickly sold out at $600 and soon began reselling in some metropolitan markets at $950. I know of one incident in which a company executive, upon seeing an ad for the JWC in the morning paper, immediately cancelled an important business conference and instead, ran down to his local sporting goods store to make sure that he would be one of the first to obtain the unique bow-levered carbine. The gun now hangs in his conference room.

Admittedly, there are some individuals who shun these newly-made "instant collectibles," with the notion that for the same price of a commemorative, an authentic 19th Century Winchester might be obtained, one which has the promise of having seen some *real* history. The sad fact is there are more history enthusiasts than there are historical Winchesters; the commemorative market was established to narrow this gap. That brings us to another important benefit of the Winchester commemorative; they attract some individuals into the gun-owning fraternity who might have never possessed a firearm. I will always remember the time I visited the home of an anti-gun individual, lured there by our common interest in model railroads. For years I had to defend myself against his verbal analysis as to why I would be attracted to "weapons of violence." You can imagine my surprise, when, upon entering his train room I spotted a Winchester Golden Spike Centennial mounted on the wall over his train layout.

"What's that?" I shouted in feigned horror, pointing to the carbine, "a weapon of violence?"

"Oh no it's not," my fellow rail fan stated matter-of-factly, "it's merely a symbol of the

The Legendary Frontiersman Commemorative Rifle was announced in 1979 and was notable in the fact that it re-introduced the .38-55 cartridge (one of the original chamberings for the 94) to American shooters.

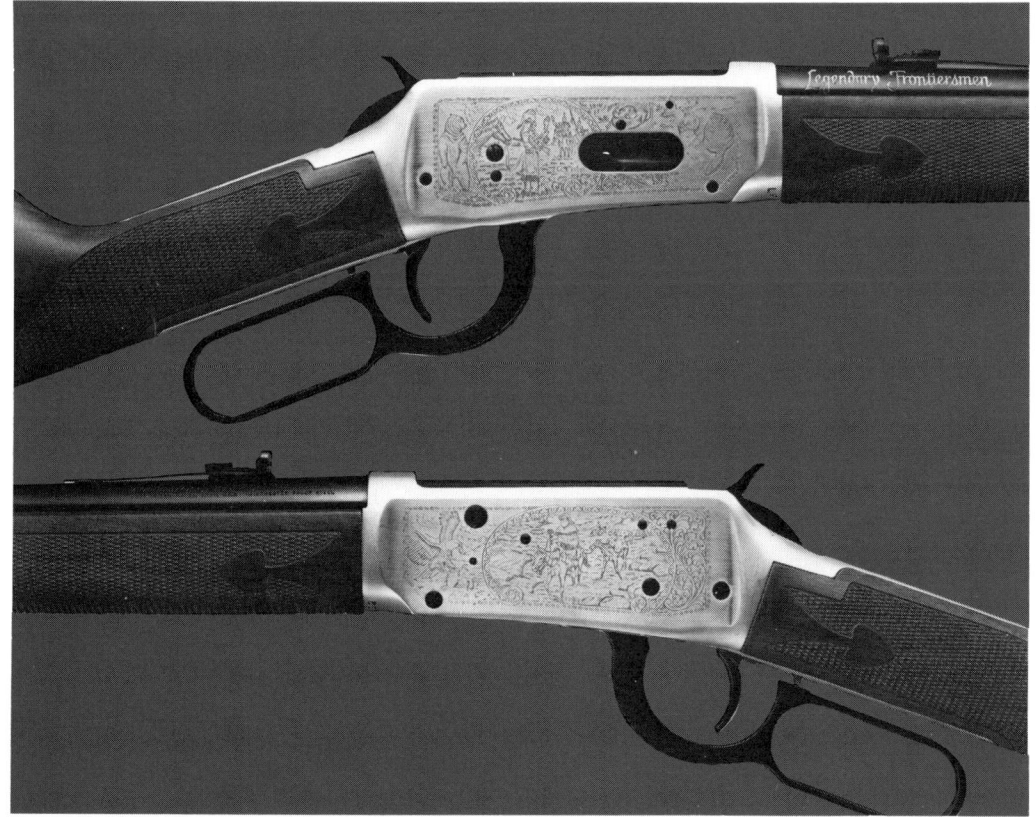

joining of the Union Pacific and the Central Pacific at Promontory Point in 1869." To him, it was not a gun at all, merely a commemorative.

Other individuals make even less of a distinction between a commemorative and a rifle. It is common knowledge that limited edition guns should not be shot. To do so severely lessens whatever value they may be destined to have. Yet I recall with chagrin an article I read in 1969 by another gun writer who decided to take the newly announced Theodore Roosevelt Commemorative rifle on a deer hunt to see how the gun would perform. Evidently it did quite well, for according to the article, he downed a

Close-up of the attractive detailing of the legendary Frontiersman. Like all commemoratives, the receivers are either roll stamped or acid etched, rather than engraved.

buck with it. Then he no longer had a commemorative; he merely had a Theodore Roosevelt rifle.

I must admit that some Winchester commemoratives simply beg to be shot. Having spent more than my fair share of hunting seasons packing short range rifles through some of the most rugged country in the Southwest, I went a little glassy-eyed when I first was shown Winchester's Legendary Lawman Commemorative at the 1978 Las Vegas Gun Show. With its short 16-inch "trapper" barrel, I was convinced that this was the gun for every backpacker and horseback rider who ever

"...It is unwise to start collecting Winchester commemoratives with the goal of getting rich quick. Rather, it makes more sense to simply purchase the gun for its own sake, as an item of enjoyment."

ventured out during deer season. I spent a full two hours trying to convince a Winchester official to produce a non-commemorative version of the Lawman. A few years later they did and I acquired one of the first ones off the assembly line. Oddly enough, I have never fired it; the Winchester Trapper is still in its original packing box with all of its factory tags intact, as it has become a personal type of commemorative for me.

It is obvious that the beauty of any of the Winchester commemoratives would be in the eyes of the beholder. Their price tags also puts their acquisition in the hands of those who are steadily employed, although some of the guns, such as the 1000 matched sets of consecutively numbered Model 94s and Model 9422s which retailed for $3,000 per set in 1979, limit themselves to collectors who can afford to soar to the higher strata of Winchester's rarities. Other commemoratives, such as Winchester's 1973 issues of the Royal Canadian Mounted Police and the Texas Ranger, or the recently introduced U.S. Border Patrol, are aimed at a specific audience, and not the mass market at all. Thus, these purchasers have a very personal identity with their guns, and it may very well be the only Winchester commemorative they will ever buy.

For the true collector the acquisition of one commemorative is merely a reason to buy another. Many of these purchasers are professional people who have already run the gamut of paintings, silver or African art. To them, the Winchester commemorative holds the portent of romance and history, even though it is largely symbolic. Investment, of course, is always the rationale for acquiring another addition to a commemorative collection, but perhaps the sagest advice comes from Terrance S. Parsons, arms & armor director for the well-known San Francisco auction house of Butterfield & Butterfield.

"Generally," says Parsons, "it is unwise to start collecting Winchester commemoratives with the goal of getting rich quick. Rather, it makes more sense to simply purchase the gun for its own sake, as an item of enjoyment. Then, if it appreciates after a number of years, so much the better. But don't buy it with the thought of selling it. Hold on to it, admire it and enjoy it for the subject matter it depicts."

To be sure, not all Winchester commemoratives have appreciated. Some have actually decreased from their original purchase price due to unfavorable economic climates. In good times, others have soared in value beyond even Winchester's most guarded dreams. There is no way of predicting which of today's new issues will be tomorrow's rarities, but Parsons, who has overseen the sale and purchase of thousands of Winchester commemoratives, offers some guidelines.

"For one thing," he says, "look for low production runs on newly announced Winchesters. Find out for sure just how many are being made. Normally the fewer commemoratives there are, the more desirable they will become as collectibles. Secondly, be sure you only buy brand new in-the-box, never before fired mint guns. Commemorative boxes and packing slips can be just as important as the gun itself and I know of some collectors who will not even consider buying a commemorative unless it has the original box and papers with it. That is one of the great appeals of buying a Winchester commemorative: they are always new guns that have never been tampered with. It is not a question of 'has it been reblued' or 'is this part original?' The commemorative collector does not have these concerns to worry about."

Because it is the largest producer of commemorative arms in the United States, when Olin Industries announced it was divesting itself of its American-made Winchester firearms division, there was a great deal of concern among collectors. However, when the investor group of U.S. Repeating Arms took over, one of the first guns they issued was the now-legendary John Wayne Commemorative. The previously mentioned success of that gun is a good indication of the future of Winchester commemoratives.

"We fully intend to stay with the commemorative program," states Charles Rhodes, director of marketing services of U.S. Repeating Arms. "It is a very vital part of our company's function. As has been our past custom, we will be announcing a new commemorative in December of this year."

This new commemorative will once again be on the Model 94 action. Although Rhodes would not publicly divulge the subject matter of their new limited edition lever gun for 1983, he admitted that it would have a physical design difference from the standard gun. "It has to be unique," Rhodes stated. "After producing the John Wayne commemorative, we have to do something equally as spectacular."

Besides, after 18 years, Winchester's commemoratives now have a tradition and history of their own to maintain.♦

The Classic
TRAP GUNS

Immensely popular in the early 1900s, the great single-barrel trap guns have fallen from grace among today's shooters. Yet the classic singles endure, lovely artifacts that can still hold their own on the firing line.

MICHAEL McINTOSH

The single-barrel trap gun evolved in response to changes in the nature of the game. The original trapshooting target was a wild bird — wood pigeons and Lincolnshire blue rocks in England, where the sport began; the passenger pigeon in America, where trap-shooting was introduced about 1830. By the late 1880s, hounded almost to the brink by habitat destruction and unregulated harvest, passenger pigeons were almost impossible to obtain in the numbers needed to supply even a modest tournament.

Trapshooters tried any number of substitutes both live and inanimate — from blackbirds, purple martins, starlings, sparrows, and even bats to discs and balls of glass, pitch, resin, paper, and clay. The diehard live-bird shooters finally settled on domestic pigeons, larger and slower than their wild cousins but worthy nonetheless.

Beginning in the mid-1890s, more and more tournaments used clay targets instead of live birds. The new targets sent ripples of change through the sport. In the old, live-pigeon days, when the rules allowed two shots at each bird, the double gun dominated. But clay targets, particularly when flung by the relatively weak traps of the time, were easier to hit than pigeons, and the rules soon disallowed the second shot.

Doubles built for trapshooting remained popular almost as long as doubles themselves were produced, but two-barreled guns were something of an anomaly in what had become a basically one-shot game. Repeaters showed promise, particularly the Winchester Model 97 pump gun

Parker Single Trap SC Grade (top) and Ithaca 4E Trap. Shotguns courtesy of Jaqua's Fine Guns, Findlay, Ohio. Photograph by Art Carter

which appeared in 1897. But no repeater is as reliable, as safe, or as a finely balanced as a break-action gun. The logical choice was a single-barrel, designed for trap and built with all the care and quality of a good double.

The first such gun offered by an American maker came, appropriately, from the genius of Dan Lefever. It was introduced in the 1905 catalog of the D. M. Lefever Company, the firm Uncle Dan founded after leaving Lefever Arms in 1901. The New Lefever Single featured the adjustable, ball-and-socket hinge joint that had been part of Lefever doubles since 1885. Like all the single trap guns to follow, it was made in 12 gauge only and fitted with vent rib and ejector.

Barrels were imported from Europe, made of steel that Lefever trade-named Imperial and offered in lengths from 26 to 32 inches. Stocks could be straight-hand, half-hand, or pistol-grip style. The single presumably was available in the same seven grades as the New Lefever doubles, though the catalog illustrates only a plain grade priced at $38.

The New Lefever was short-lived. Uncle Dan died in 1906, and the D. M. Lefever Company folded soon after. There is no evidence that Lefever Arms, which remained in business until 1915, ever built a single-barrel trap gun.

The first single to have a significant impact on the American industry came from a company founded by William Henry Baker. Like most of the guns that bear his name, Baker's single is not as famous as it ought to be.

William Baker played a key role in the early development of the American shotgun — as the founder of both the L. C. Smith and Ithaca Gun companies, and as the engineering brains of the Baker Gun and Forging Company. The most historically important of the Baker guns, however, was not an original William Baker design, but rather an adaptation.

Baker died in 1889, leaving the company in the hands of his brother Ellis. By the turn of the century, the firm was deeply involved in trap-shooting. It maintained a shooting grounds, hosted tournaments, and acted as principal sponsor for W. R. Crosby, one of the greatest pigeon shots of the day. As the game evolved from live birds to targets, Baker recognized the new market and in November 1909, announced the birth of a single-barrel gun built specifically for the target-shooting trade.

Actually, the gun was delayed by last-minute design changes and did not appear on the market until 1912. The bolting system was to have been the sliding underlug used in Baker doubles, but the design staff decided that a bolt near the top of the breech would prove more durable than one at the bottom. The Baker single is fastened by two small lugs that slide horizontally out from

The first of America's great single traps came from the genuis of Dan Lefever. Production of Lefever singles was cut short by the death of Uncle Dan in 1906, about one year after the gun was introduced. A plain-grade single sold for $38.

the standing breech into recesses in the barrel, one on either side of the chamber.

The Baker came in three grades: Sterling, which sold for $65; Elite, priced at $100; and Superba, at $160. All were built to order. Unlike most automatic-ejector systems, with the mechanism fastened to the fore-end iron, the Baker ejector was fitted into the frame. For aesthetics, the rear edges of the frame were deeply scalloped, which added a degree of grace to the tall, slab-sided profile. Later Sterling Grade frames were not scalloped, but sculpted with a single flute along the rear of the barrel, extending nearly to the rear of the frame.

The Baker design was revised about 1916. The right-hand locking bolt was eliminated, the ejector system changed to a more conventional — and considerably less durable — design, with the works in the fore-end. The integral rib was replaced by a higher, free-floating type.

These changes were prompted by economics. Like its peers in the sporting arms industry, Baker's fortunes were devastated by World War I, and in 1919, it sold the gun business to the H & D Folsom Arms Company of New York City. The old, high-quality Baker line, the single traps included, came to an abrupt end. The original records are long gone, but it seems likely that only about 2,000 Baker singles were built.

That Bakers do not enjoy the reputation they deserve is largely a quirk of history. Unlike the double gun, which in America arrived at the peak of its quality in the years just before the Great War, the single trap found its Golden Age in the 1920s. By then, the great ones — Ithaca, L. C. Smith, Parker, and Fox — were in full bloom, and the Baker was little more than a memory.

Emile Flues designed the first Ithaca single. He was a free-lancer who developed the last of the old-style Ithaca doubles. Like them, his single-trap action is fastened by a sliding underlug in combination with twin lugs at the breech-end of the barrel and a cross-pin bolt.

Four prototypes, serial numbers 246892, -3, -4 and -5 were built in 1914. When full-scale production began later that year, the Ithaca single was offered in the same grades as the double guns, No. 1 through 7. The plain Victory Grade was added in 1919.

An even higher grade than No. 7 appeared in 1916, created especially for the single trap and named after composer-bandmaster John Philip Sousa, who at the time was president of the Amateur Trapshooters Association. The Sousa Grade is lavishly engraved, with dogs, ducks, and even, at Sousa's request, a buxom mermaid inlaid in gold. In the first year, it cost $500.

The Flues single became obsolete in 1921, when Frank Knickerbocker designed what would be the most famous and enduring single trap gun in America. It was called the Knick, and the first one, serial number 400000, left the factory on May 1, 1922, bound for Boston.

Knickerbocker did with Emile Flues's design what Flues had done with those before him — scrapped it and started over. The Knick hammer and sear, placed in the center of the frame, are powered by a vertical coil spring that bears directly against the hammer. The cocking system is a rod-and-cam arrangement. The action is fastened by a double-wedge bolt near the top of the standing breech — similar in construction and identical in principle to the Baker system.

The Knick was an immediate success. By 1935, Ithaca had assembled an impressive number of Grand American Handicap winners around the faithful Knick — A. E. Sheffield, Charles Young, Mose Newman, Elmer Starner, Charles Larson, and John Henry.

Unveiled in 1912, the Baker single-barrel trap gun was little more than a memory by the 1920s — the Golden Age of single guns. But the great ones — Ithaca, L.C. Smith, Parker, and Fox — were in full bloom.

There were even more famous fans. Annie Oakley shot an Ithaca single, both in competition and in exhibitions. Sousa's favorite trap gun, naturally, was a Knick. In time, Generals Dwight Eisenhower and George Marshall received ornate presentation guns.

From the start, the Knick was available in all grades from Victory to Sousa. The No. 6 was discontinued in 1919 and the Victory Grade in 1937. The Sousa Grade was renamed $1000 Grade in 1937, and from the end of World War II, the name changed every few years and in increments of $500. By 1980, it was simply the Dollar Grade, built on special order.

When Ithaca bought Lefever Arms in 1915, it discontinued all of the old, Lefever-designed guns, but kept the name for a series of inexpen-

sive guns that first appeared in 1921. A single-barrel version, brought out in 1927, was available as a vent-ribbed trap gun. It enjoyed some modest popularity, but in the high-quality trap-gun world, "Ithaca" meant the Knick.

Thanks to William Brophy, who scoured original factory records before writing his book *L. C. Smith Shotguns*, we know a great deal about the production of these guns. According to those records, the Smith Single-Barrel Trap Gun was first produced in 1917.

The Smith single, unlike the double, is a boxlock, but the design incorporates two of the features that made the Smith doubles great: the rotary-bolt fastener and the unique, crank-type cocking system designed by Alexander Brown.

Strong as it is, a rib extension is a difficult concept to apply in a single-barrel design. There is plenty of room to fit it between the barrels of a double and still maintain a shallow frame; in a single, it must either be slimmed down considerably or moved to one side of the breech. In either case, the barrel walls must be left quite thick at the breech end. The Smith designers reduced the depth of Brown's rib extension by about half and left it at the top of the barrel, along with enough steel to accommodate it.

Brown's cocking system, designed in 1886 as part of the first hammerless Smith double, adapts quite well to the single. The cocking rod has a cranklike lever at one end, a cam at the other. The crank fits into a recess in the fore-end iron; as the barrel is lowered, the rod rotates and the cam lifts the hammer into its sear notch. There are simpler designs and probably stronger ones, but you'll look a long time to find a Smith with a broken cocking rod.

In the first couple of years, the Smith rib was free-floating. Though company advertising made much of that, the design was changed in 1919 to a rib milled from a single strip of steel.

When it first appeared on the market, the Smith single was offered in four grades — Specialty, Eagle, Crown and Monogram — at prices ranging from $100 to $310. Barrels could be 30, 32, or 34 inches long. A rubber recoil pad and twin ivory bead sights were standard fittings. At customer request, the trigger could be in the center or rear position.

Like the others of its kind, the Smith proved popular, and company advertising presently began counting coups of tournaments won. In 1926, Miss Belisa Gleaves won the Virginia State Amateur Trap Shoot with a Smith single and a score of 192 x 200. She was the first woman ever to win a state shoot.

After the rib change of 1919, no further revi-

Early Grand American Trapshooting Championships, held in Vandalia, Ohio, drew huge crowds. Single traps were favored by many American shooters after World War I.

sions were made in the Smith design, and the remainder of its production history is marked only by new grades added or old ones discontinued. Between October 24, 1927, and May 25, 1928, a dozen singles were built in Whippet Grade — the only Smith guns of that grade. Factory records show that all had 32-inch barrels but offer no clue to either the nature or extent of their engraving.

All L. C. Smiths, single and double, were discontinued in 1950. Thirty-three years of production accounted for only 2,666 single traps, 1,861 of which were Specialty Grades. Though it was never listed as a catalog item in more than six grades, Brophy's research turned up singles built in every Smith grade. Except for Specialty and Olympic, fewer than 100 were built in any one grade, fewer than 20 in most.

For so famous a gun, surprisingly little research has focused on the Parker single trap. We do know that it first appeared in 1917. Typically conservative, Parker saw an opportunity to improve, at least cosmetically, on an established market; if the guns represent no particular mechanical advantage, they certainly offered the customer a chance to own the most lavishly-finished trap gun of the day. Grades begin at C and go all the way to A-1 Special, which in 1917 cost the goodly sum of $550.

This isn't to imply that the Parker single is all floss. To any firm accustomed to building high-quality doubles, a single trap is no particular challenge. All it takes is half of the double's action and ejector system, a much-simplified single trigger, and some creativity in designing a way to bolt the barrel to the frame. Parker designers chose the time-tested underlug, which makes their gun the only great American single to be so fastened.

The Parker single became extremely popular among trapsmen. Fred Gilbert, one of the great live-pigeon shooters who proved equally skillful at targets, used one to set a new world's record for registered clays — 569 straight — in 1920.

Once in production, the single's design was never changed in any substantial way. When Parker sold out to Remington Arms in 1934, the single remained in production, renamed the Parker Model 930. Remington discontinued all Parkers in 1942. No records survive to show the number of singles built, but since Parkers greatly outnumbered any of the other great doubles, it seems likely that the singles did, too — perhaps as many as 10,000 of the nearly quarter-million Parkers built.

World War II put a major kink in the A. H. Fox Gun Company's plans to claim a share of the single trap market. The gun was still in development and testing stages in 1917, and the whole project ended up in limbo while Fox struggled to survive the wartime economy. It finally was put into production in 1919, the last of the great American singles.

Rather than adapt the famous Fox rib extension and rotary fastener to the single, the Fox designers instead chose a Greener-type cross-bolt that engages heavy lugs on the barrel. The barrel itself is made of Chromox steel, Fox's own formula, developed about 1910. The factory put great care into boring and polishing barrels. Chokes, too, bored with what Fox liked to call "straight-taper" cones, were rigorously honed and tested for dense, even patterns.

Ithaca's Sousa Grade was lavishly engraved with dogs, ducks, and even a buxom mermaid inlaid in gold. In its first year it cost $500.

The Fox has one unique feature — a top-latch release on the outside of the frame. In the old days of leg o'mutton and trunk-type gun cases, it was customary to store and transport guns broken down. They took up less space that way, and the heavy leather cases protected them better than any full-length soft case. The only problem is that when you dismount the barrels from virtually any break-action gun, the top-latch stays in the open position until you trip the catch that holds it. The trip usually is in the water-table slot or the rib-extension slot, reachable only with a key or screwdriver. Getting to it is no particular hardship, but putting the release button outside the frame is typical of Fox thoughtfulness and attention to detail.

From the beginning, there were four grades: J, K, L, and M. Their decoration corresponds roughly to the C, X, D, and F grade doubles, respectively. Catalogs issued before 1930, however, note "Special High Grade Guns built to order," so it's possible there's a Fox single or two more ornate than M. Grade.

Like Parker, the Fox company was on the ropes by the end of 1929. Savage Arms bought all the machinery, patents and inventory in 1930 and tried mightily to keep the Fox afloat during the '30s. It was an uphill struggle. By 1935, only J and M Grades remained. Though Savage records indicate that the last Fox single left the factory on September 13, 1935, the gun remained in their catalogs until 1940 — in J Grade only, after 1937. By 1942, all Fox guns were gone for good.

Over its 16 years of production, Fox made only 568 singles, an average of 33 guns per year. Only the D. M. Lefever singles are scarcer.

The great old single traps have by now fallen from grace among the most serious shooters, supplanted by high-tech pieces built from better steels, fitted with faster triggers, and decked out with choke tubes, ported barrels, and adjustable ribs. Some even look like trap guns. Still, the old ones endure, lovely artifacts that can still hold their own on the firing line.

Around the central-Missouri countryside where I live, Sunday-afternoon turkey shoots are popular. Targets are three-by-five index cards with a X drawn in the middle; the prize goes to the gun that will put the most pellets nearest the cross at 40 yards, using factory shells supplied by the shoot sponsor. It's almost wholly a test of barrel and choke. So far as I know, only one gun is no longer allowed to compete. It belongs to a friend of mine, and it won every contest until the sponsors simply disqualified it. They'll still let Doc shoot, of course — but not with his old Fox single. ♦

Ithaca's 5E Single Barrel Trap. This vent-ribbed design enjoyed great popularity after its introduction in 1927.

MODEL 12
The Perfect Repeater

Often referred to as the greatest repeating shotgun ever built, Winchester's Model 12 was produced in every conceivable gauge and grade — more than two million guns in all.

DAVID E. PETZAL

Picture in your mind's eye a Yankee grouse hunter, his setter locked on point at the edge of an ancient apple orchard. He advances on the dog, dressed in weathered, briar-torn canvas. What shotgun is he carrying? Obviously, a Parker or L. C. Smith.

Or, let us travel miles south and envision a quail hunter riding in a mule-draw wagon through the scrub pines that bobwhite call home. Ahead of him ranges a pair of pointers, and cradled across his lap is . . . a Purdey, Boss, or Holland & Holland.

Do these gentlemen hunters carry pump guns? Not on your life. The fine double is the glamour gun, the one that collectors kill for; but the repeater? Well, it's a good using gun if you can't afford something better, but it's really just a collection of parts that fires shotshells. The pump gun is the private first class among scatterguns, the one detailed to pull KP, do guard duty and dig slit trenches.

There is, however, one exception: the Winchester Model 12, the paragon of pump-action shotguns.

The two-barreled shotgun is a European form, which was already highly evolved when the 13 American colonies fought for independence. On the other hand, the pump gun — indeed, repeating arms in general — is purest Yankee. It was designed for men who wanted to put meat in the pot, aesthetics be damned, and who could not afford to spend much to do it.

The origin of the double-barreled shotgun will forever remain a mystery, but we do know that the pump or slide-action shotgun was the brainchild of Christopher C. Spencer, inventor of the Spencer carbine, which was loved by the Union calvary and loathed by good Confederates everywhere.

Like the rest of Spencer's civilian efforts, the Spencer pump (which appeared in 1884) was not a commercial success, but the idea did find a warm welcome at Winchester Repeating Arms. Winchester's first slide action was the Model 1893, a John M. Browning design. It did well enough but could not handle the pressures developed by smokeless-powder shotshells, which were just then becoming popular.

Winchester phased out the Model 1893 and reissued it, with an exposed hammer and a strengthened frame, as the Model 1897. Complicated, ugly, incredibly durable and hugely successful, the Model 97 was in production until 1957 — a life span of 60 years. Waterfowl hunters doted on it. The 97 was a long arm with most of its weight forward, and once you started swinging it, it was hard to stop, a fact which caused many a mallard to end up in a game bag.

Over a million Model 97s were made, but it was still a flawed gun. It was complex. When you work the action, you are aware of a huge number of parts being put into motion. (I knew a village constable who preferred the 97 precisely for that reason. "When people hear that thing being shucked in the dark," he said, "they change their minds about whatever they're about to do.") The exposed hammer was also a problem. It was small and powered by an extremely muscular spring, which made it difficult to control with a cold, wet thumb — or any other kind of thumb, for that matter. It could also bite a chunk out of your finger as efficiently as any machine devised by man.

The popularity of the pump gun was established beyond a doubt, but there was room for something better. That something was the Model 12.

Unlike the Model 1893, the Model 12 was not the result of John Browning's genius; instead, it resulted from the team effort of Winchester's designers under the direction of Thomas C. Johnson. In all likelihood, Johnson did not come up with the idea for the gun by himself, but he must take much of the credit for what has been called the greatest repeating shotgun ever built.

The Model 12 was an astonishing departure. It was first offered as a 20-gauge with a 25-inch barrel. (Two years later, in 1914, 16- and 12-gauge versions were also introduced.) Where the 97 was lumpy and angular, the Model 12 was sleek. The exposed hammer had been transferred inside the frame, and the action vastly simplified. It was as durable and reliable as the 97 — probably more so — but it was comparatively light, quick-handling and graceful as well.

However, like so many of Winchester's great designs, its own excellence assured its eventual demise. The Model 12 was built with the philosophy that said, in effect: "We'll make the best gun we possibly can; manufacturing costs will have to come second."

The Model 12 owed its existence to the fact that first-class machinists worked cheaply in those days. The mechanism was not complicated, but the receiver, which was machined from a solid steel forging, was a nightmare. David F. Butler, in his book *The American Shotgun*, describes the travail involved in making a Model 12 receiver:

"The locking recess, for example, was so difficult that a two-piece machining setup had to be used. An extension drive was pushed in from the front of the receiver at an angle to engage a separate cutter assembly which was inserted from the bottom to machine the locking recess. Another reason for the high expense can be seen in the chamber construction. The shotgun was designed with a takedown system which allowed removal of the entire barrel assembly. In order to keep headspace within tolerance, a separate ring was fitted to form the rear section of the chamber. Special machining operations were required to assure concentricity between this separate ring and the remainder of the chamber.

"Model 12 shotguns were designed almost without regard to cost. Nickel-steel barrels were fitted and chrome-molybdenum alloys were used in receivers and interior components. This meant that the gun was always very high-priced in relation to other slide-action shotguns. It also meant that the guns were virtually indestructible and if given adequate care and cleaning would last several lifetimes."

The Model 12 was called "The Perfect Repeater" with reason. Its durability was almost beyond belief. Model 12s were made in many configurations and gauges and for every kind of game that flew; they were cobbled into riot guns; they were made plain and they were made fancy. Of all the places where they were used and loved, however, it was on the trap and skeet field that the Model 12 was king.

There is a proverb among automotive engineers that you don't really know what a car can do until you race it. On the track, there are stresses that are impossible to duplicate under any other conditions. Similarly, with shotguns, trap and skeet shooters can best tell you what kind of a gun you have. Serious shooters of these sports go through shells in carload lots, and a gun that is less than dead-reliable is intolerable. One season of serious campaigning has reduced many a gun — splendid though it was in other use — to rubble. The Model 12 was one of the few guns that could stand the gaff, then and now.

In addition, it has a quality that is supremely difficult to design into any gun, much less a pump: it balances and points splendidly. There is just enough forward weight to keep you swinging, but you never feel that there is a dead weight in your hands. Look down the rib of a Model 12 trap gun, and you'll find it hard to imagine yourself missing. Heft the sleek, deadly weight of the Model 12 Duck gun with the 3-inch chamber; you can almost see the geese going limp in the air. There have been few firearms like it.

However, not even the Model 12 is perfect. If you hold the trigger back and push the fore-end forward, the hammer will fall and the gun will fire. If you didn't know this and were careless about where you pointed the muzzle, you could kill someone. (This defect was corrected in later 12s, but be warned.)

Experienced shooters could control this slam-fire characteristic and turn it to their advantage. A well-used Model 12 and a skilled shot could, by pumping the fore-end hard and fast, get off a series of shots about as quickly as a man with a semi-auto. Double targets at skeet and trap were no problem for Model 12 shooters. Herb Parsons, one of the greatest exhibition shooters of modern times, used a Model 12. One of his favorite tricks was to throw seven clay targets in the air at once and break all of them.

Like so many of Winchester's great designs, the Model 12 was built with the philosophy that said, in effect: "We'll make the best gun we possibly can; manufacturing costs will have to come second."

Slam-fires notwithstanding, the Model 12 enjoyed a long and illustrious life. Despite its high price and competition from other excellent repeaters, it remained in the Winchester line until December 1963, at which point its high manufacturing cost forced the company to relegate it to the status of special-order only. At that point, 1,968,307 had been produced.

So great was the demand for the Model 12, however, that in 1972 Winchester reintroduced the gun, in 12-gauge only, into its regular line in trap, skeet and field models. But the handwriting was already on the wall, and four years later the skeet and field models were discontinued. The trap gun hung on until 1980, when the days of the Model 12 were finally over. At that point, a total of 2,026,721 had been produced.

What killed the Model 12 was not so much its price but the changing tastes of shooters. Trap and skeet competitors had learned that the choice of chokes offered by an over/under have a major competitive edge and that the recoil reduction made possible by the gas operation of the Remington Model 1100 auto could reduce their tendency to flinch, walk in circles, and speak in Urdu after a string of several hundred birds. In the final analysis, it was obsolescence rather than price that eliminated the Model 12.

This is not to say that the Model 12 is not a viable gun. It is still a splendid piece of machinery and will perform quite nicely for you however you decide to use it.

If you would like to invest in a Model 12, beware. There are more than two million of them around, in every conceivable gauge, grade, and type. As with any other highly popular firearm, everything that could be done to the Model 12 was done at one time or another, by owners, gunsmiths, the Winchester factory, and unfortunately, by some bozos and clods. Some Model 12s were beautifully engraved and customized, and they are now valuable guns. Others were butchered; they are useful only for spare parts.

To give you some idea of the variations in Model 12s, look at the following:

The Model 12 was made in 14 different styles and in gauges from 12 to 28. Barrel lengths ranged from 26 to 32 inches and could be had in chokes ranging from full to improved cylinder, with all steps in between.

You could purchase a Model 12 with a solid or vent rib and with any style of stock you chose, appropriate to the type of gun. In 1964, some competition guns were even produced with Hydro-Coil stocks, which were made of plastic and employed internal pistons to delay recoil. (They were also unspeakably ugly.)

The Model 12 was produced in an era when the Winchester Custom Shop would make anything you wanted for a price. As a result, there exists a considerable number of custom Model 12s that bear little relation to factory specs.

As with other shotguns, some fairly standard rules apply to determining the worth of a Model 12. First, obviously, is condition. Many of these guns were used hard, and many of them were abused. A Model 12 in poor or even fair shape is not much of an investment unless it is extremely rare.

As a rule, the smaller the gauge, the higher the price. There were fewer 28 gauges produced than any other, and this is reflected in their value. For example, it is reasonable to ask $2,500 to $3,000 for a 28-gauge field gun in excellent condition.

Pedigree is important. A Model 12 in the factory box with all its tags is highly desirable, and factory records of custom-made guns, as well as a history of ownership, will make a collector salivate. Guns that remain intact and unaltered are hard to come by because so many people worked on them, and a 12 that is reblued, refinished, or restored in any way at any place other than the Winchester factory is worth less than one that is "as issued." A reputable seller will specify, whether in an ad or in person, exactly how much original finish is left and whether the gun has been reconditioned. If the owner of the gun doesn't provide the information, pass it up.

If you are considering a highly decorated Model 12, fixing a value depends on whether the gun was engraved and stocked at the Winchester factory or somewhere else. As a rule, factory-fancy 12s are considerably more valuable, unless you are considering a 12 that was engraved and/or stocked by craftsmen of major stature. If you come across one that was stocked by Tom Shelhamer and engraved by Josef Fugger, for example, it will be worth quite a bit of money. Gary Herman of Safari Outfitters, Ridgefield, Connecticut, says the most expensive Model 12 he could conceive of is a full-fancy 28-gauge factory gun from Winchester's "golden era" of 1920-1940, accompanied by authenticating paperwork. It would bring about $16,000.

Prices for lesser 12s are more modest. Competition guns in excellent shape run about $800, and field guns in the same condition will run from $400 to $700. If you are looking for a "user," you are advised to take it to a gunsmith and have it checked for tightness and correct headspace. The Model 12 is an extremely easy gun to work on, and repairs tend to be simple and inexpensive. (You'd better decide why you want a Model 12 skeet gun if it is fitted with a Cutts Compensator, as many of them were. The Cutts was an effective muzzle brake and allowed for interchangeable choke tubes, but it produced a muzzle blast to the side that is distinctly unwelcome. If you want to collect such a gun, fine, but if you plan to use it, think twice.)

Before you spend a cent, however, you must read, talk, and investigate. Talk with dealers who specialize in collectible guns: Buckhorn Quality Firearms — (214) 221-8583, Chadick's Ltd. — (214) 563-7577, Jaqua's — (419) 422-0912, and Safari Outfitters — (203) 544-8010. Check out the prices and the guns themselves. Then balance both against what you can afford.

Look through a couple of issues of *Shotgun News* or a new publication called *The List* (which is similar to *Shotgun News* but lists guns by model, thanks to the miracle of computerization). Get a feel for what guns are available and what they cost.

Finally, remember that you are unlikely to make a killing in this market. Legions of Model 12 collectors have already scoured the barns and attics of America looking for guns owned by widows who sell them for $35. If you see something sensational that is selling at a bargain price, the warning signals should go up. At worst, it could be a stolen gun, which is real trouble.

No machine made by the hand of man is perfect, but the Model 12 came closer than many. If you buy one, it will be your good friend and probably your great-great grandson's too. ♦

PATRIARCH
of the Gun Shops

It's been 64 years since Seymour Griffin and James Howe opened their Manhattan rifle shop. And while their firm has changed location several times, it has never lost sight of its concern for the highest levels of craftsmanship.

MICHAEL McINTOSH

Midtown Manhattan, New York City, seems an odd place to find a gunshop. In the teeming, noisy canyon of West 44th Street, you might as readily expect to see a trout stream or a skeet range. But if you walk west from Fifth Avenue, you'll find the Bar Building, No. 36, midway down the block. In through the double glass doors, and most of the noise drops away behind. Down the long, marble-arched entryway to a rank of elevators, interiors paneled in milled oak. Off at the tenth floor, into an anonymous hallway where the opposite wall wears a shiny brass plate stamped with two names above the familiar stylized head of a bighorn ram and a discreet arrow routing left.

You'll know the place when you turn the last dogleg in the hall and go through the big oaken door at the end, past another polished brass plate that reads Griffin & Howe, into a world at a crossroads of time.

Custom-made .35 Whelen. Photography by Art Carter.

One road reaches back, a long way back, as American custom gunshops go. At 64, Griffin & Howe is the patriarch.

It began with Seymour Griffin, a young New York cabinet-maker fond of rifles. In 1910, as the story goes, he read Teddy Roosevelt's *African Game Trails* and was intrigued by TR's praise of the Springfield rifle. Owning no rifle of his own, Griffin bought a blank of French walnut for $5 at Von Lengerke & Detmold and married it to a military Springfield barreled action. Over the next dozen years he continued to stock Springfields in his spare time.

Somewhere along the way, he met Col. Townsend Whelen, the great rifleman and technical writer. In 1923, when Whelen was officer in charge at the Frankford Arsenal, he introduced Griffin to James Howe, a Frankford gunsmith with an impressive talent for metalwork. The two men struck a bargain, if not a friendship, and opened for business June 1, 1923, at 234-240 East 39th Street, New York, under the style of Griffin & Howe.

They offered the right thing at the right time, for the bolt-action sporting rifle was just then coming into its own. It had already proven itself as a military weapon during the Spanish-American War, much to the discomfiture of the U.S. Army. At San Juan Hill, Cuba, in 1898, 15,000 American troops carrying trapdoor Springfields had been shot nearly to dollrags by only 700 Spaniards armed with Model 1893 Mausers. The '03 Springfield was developed as a result, a gun designed so closely along Mauser lines that the U.S. government had to pay Mauser $1 million in royalties. But it was only after American soldiers came home from World War I, experienced at using Springfields and bringing with them captured Mausers, that the bolt-action really caught the hunter's eye.

When Griffin and Howe set up shop in 1923, none of the American arms-makers offered a bolt-action sporter. Around the turn of the century, Winchester had built about 1,700 Lee Straight Pull military rifles in a sporter version, but they were long out of production. The Winchester Model 54 would not appear until December 1925. Before that, the hunter who wanted a bolt-gun had to get it from England, Germany, or one of only a handful of American custom shops. R. F. Sedgley of Philadelphia had been turning out low-priced Springfield sporters since 1911, but with stock work of only middling quality, Sedgley never posed much competition to Griffin & Howe. Louis Wundhammer, working in Los Angeles, had been at it even longer than Sedgley and may in fact have built the first Springfield sporter of all, for writer E. C. Crossman in 1910. But while Wundhammer's quality was high, his output wasn't.

Hoffman Arms, which opened for business in Cleveland the same day Griffin & Howe started in New York, proved the only real rival. In December 1923, James Howe left the partnership and went to work for Hoffman. It probably was not an amicable parting, but his name remained. Seymour Griffin was able to secure a number of European gunsmiths, mainly Germans and Austrians, and set to work in earnest. Griffin & Howe found its essential, if not its literal, birth in those early months of 1924 when by dint of sheer effort, long hours, and an insistence upon only the highest level of craftsmanship, the firm began to flourish.

James Howe went on to leave his own mark. His book, *The Modern Gunsmith,* is still a standard reference in the trade.

The national economy prospered in the 1920s, and each year more well-heeled American sportsmen set off to hunt big game in North America and faraway corners of the world. By the middle of the decade, a growing number of those who could afford the best custom work were of a mind that the best was available at Griffin & Howe. The shop turned out 820 rifles from mid-1923 through the end of 1929. As patronage among the carriage trade increased, Griffin moved to larger quarters on East 44th Street and opened a retail store to go along with the custom shop. By the end of the '20s, Griffin & Howe was a complete outfitter, purveyors of everything from boots and safari clothing to canvas tents and collapsible camp furniture. It even sold ammunition under its own brand name.

Superb metalsmithing and classically elegant stock work had always been Griffin & Howe hallmarks, but now other distinctive features started to appear. When telescopic sights began to interest American hunters in the 1920s, Griffin & Howe designed and patented one of the first successful side mounts. It not only held a scope securely in a position comfortable for the shooter but also, through an ingenious dovetail-and-lever arrangement, allowed the scope to be dismounted and then replaced without wrecking its adjustment. It's still available. So are the Griffin & Howe hooded ramp front sight and the barrel band for attaching a front sling-swivel well ahead of the fore-end, which allows a rifle to ride lower on the shoulder and the barrel to catch fewer snags in heavy cover.

Besides the stock design, the most distinctive Griffin & Howe detail probably is the famous quarter rib. Originally designed as an Express-type rear-sight base, this is a milled-steel rib that extends over the front receiver ring and about six inches down the barrel. It typically includes a dovetailed sight assembly with a standing bar

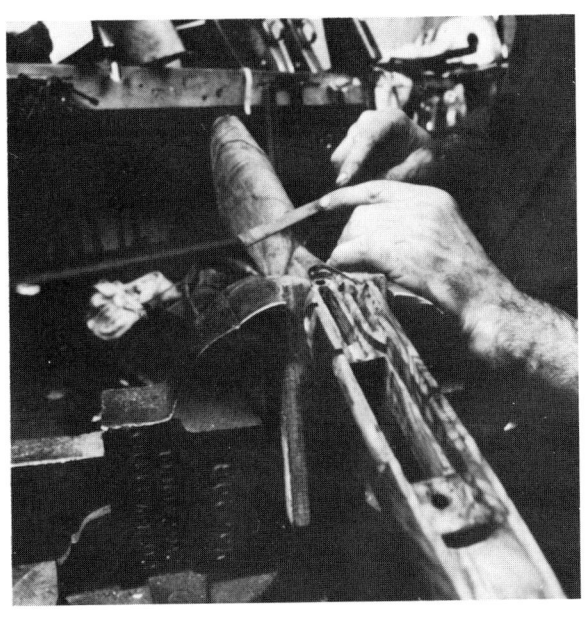

Superb metalsmithing and classically elegant stock work are hallmarks of Griffin & Howe.

and as many as three folding leaves calibrated for various ranges. It can be fitted integrally with the front scope mount and can be installed with or without the open sights — but if the quarter rib isn't there, it won't look like a proper Griffin & Howe rifle.

Hunters in the 1920s soon recognized a need for new, more efficient cartridges, especially in the middle calibers. Most of those available from the ammunition industry had originated in black-powder days and were, by the end of World War I, woefully obsolete. The most productive period of experimentation and wildcatting was just beginning, and Griffin & Howe played a part.

The gun shop is a Dickensian clutter that every first-rate workshop ought to have — benches scarred and splotched with gun oil, the smell of walnut dust and linseed oil in the air.

By far the best of the mid-bore cartridges available then was the great .375 Holland & Holland Magnum, introduced in 1912. But it required a long action, and the only one available was the Magnum Mauser, hard to come by and expensive besides. About 1922, James Howe designed an excellent alternative by necking up a .30-06 case to accept a .35-caliber bullet. He named it for his old friend Col. Whelen. Aught-six cases were plentiful; the cartridge was, of course, perfectly suitable for a Springfield or standard Mauser action; and the .35 Whelen became extremely popular. It could handle any North American game and all but the largest African and Asian animals. No commercial gun-maker ever put it into standard production, but the custom shops, Griffin & Howe included, turned out .35 Whelen rifles by the dozen. You can find one even now without much trouble, and if you're a handloader, it's still a first-rate choice.

The one cartridge to bear the Griffin & Howe name appeared about 1927 — a .375 H&H case necked down to .35-caliber. Although it originally was called the .35 Holland & Holland Magnum, Griffin & Howe designed it. Eventually it came to be known as the .35 Griffin & Howe Magnum and even later as simply the .35 Magnum. It wasn't a bad round, especially for really big game, but it wasn't as good as the .375 nor as efficient as the various .35 short magnums and, naturally, could only be built on a long action. It never stirred up much interest.

But the 2R Lovell did and while not the inventor, Griffin & Howe did more than any other gun-maker to support what probably was the most popular .22 wildcat cartridge ever developed. Actually, it was the product of compound wildcatting. About 1934, Hervey Lovell designed a .22 cartridge based on the .25-20 Single Shot case, which had first appeared in 1882 as a wildcat. Then about 1937, Harvey Donaldson redesigned the .22 Lovell (also called the .22-3000) as the 2R Lovell, which proved to be a superbly accurate round in both bolt-action and single-shot rifles. The varmint-shooters loved it. Griffin & Howe built dozens of rifles and custom-loaded the ammunition by the bushel. When the supply of .25-20 cases finally ran out, the company cut a deal with Winchester to make up several thousand 2R Lovell cases and continued to sell them well into the 1960s, by which time the factory-loaded .222 Remington had made the old wildcat completely obsolete.

The onset of the Great Depression jarred Griffin & Howe as hard as it did everyone else — perhaps even harder, since most of its customers were wealthy men who found their buying power wiped out almost overnight. So small a company could scarcely have hoped to survive alone. Griffin's first move, for reasons now unknown, was to take on a partner, Harry Hobbs of West Sayville, Long Island, presumably a metalsmith as Howe had been. Company records show that he bought in on June 23, 1930, and was put on salary at $75 a week. At the same time, the name was changed to Griffin & Hobbs.

Neither the arrangement nor the name lasted long. On October 7, 1930, Abercrombie & Fitch bought a substantial interest in the firm. Hobbs left, the style became Griffin & Howe once again, and the best days of all lay ahead.

Abercrombie & Fitch was about to become the most famous sporting-goods store in the world, the kind of place the word *emporium* truly fits. Its nine floors were packed with such an assortment of goods that a hunter, fisherman, camper, explorer, or adventurer bound for anywhere on earth could outfit himself to the last detail without ever leaving the corner of 45th Street and Madison Avenue. Many of Griffin & Howe's customers also patronized Abercrombie & Fitch, so the relationship was a natural. The symbiosis proved ideal, expanding Griffin & Howe's retail trade and at the same time giving Abercrombie & Fitch part ownership of the best custom gunshop in the country.

To that time, Griffin & Howe had been primarily a rifle shop, both in reputation and in fact. The hard times of the 1930s, however, brought big-game hunting, especially the expensive African and Asian jaunts, almost to a standstill. The fearsome droughts of the early '30s did much the same to North American waterfowling, and hunters took a decided turn toward the uplands.

With interest shifting to shotgun sports and with the scope of its retail potential broadening, Griffin & Howe gave a greater share of attention to the shotgun. According to company records, an unnamed gun-buyer went to England in 1930 and came back with what amounted to Griffin & Howe's starting inventory under Abercrombie & Fitch: ten matched pairs of sidelock guns, a dozen individual sidelocks, a 20-gauge try-gun, a pile of fine European walnut stock blanks, a pocketful of gold letters for inlaid initials, snap caps, cleaning rods, oil bottles, and other miscellaneous truck. There were Churchills, Blanches, Bosses, Evanses, and others — in all, 32 best-quality London guns. The invoices for the entire purchase, import duties included, came to $9,900.

Naturally, the shop continued to turn out lovely custom rifles built on Springfield, Mauser, Krag, and other actions, but production during the '30s was only a bit more than half what it had been in the previous decade, and it continued to decline. A foundering economy followed by a global war had much to do with that, but domestic competition also played a part. After the war, both Winchester and Remington offered good bolt-action rifles, which unquestionably diverted a share of the American gun market from the custom shops. Griffin & Howe built only about 30 rifles per year throughout the 1950s.

Meanwhile, though, the shotgun trade blossomed, and by the end of World War II,

Griffin & Howe was not only the most prestigious custom shop in the country, but also one of the largest retail gun-sellers in the world.

Its customers certainly included a covey of high-ticket dandies who never pulled the triggers of the splendid guns they bought, but Griffin & Howe always was at heart a shop for shooting men, famous, nearly famous, and obscure: Grancel Fitz, Ernest Hemingway, Gary Cooper, Clark Gable, Robert Ruark, Elgin Gates, Artie Shaw, Dwight Eisenhower, and legions of others.

Life in a gunshop is not without its moments. There was a 30-meter shooting range in the basement of the Abercrombie & Fitch building, where any customer was free to try out a rifle or two. On their way to Africa in the mid-1950s, Ernest Hemingway and some crony came in to browse and buy a few odds and ends. Hemingway noticed a .577 Westley Richards double rifle on the rack and told his friend it was just what he needed for elephant. A salesman gathered up some ammunition, and they trooped down to the basement.

Now, the .577 Nitro-Express cartridge is about the size of a baseball bat and many times more lethal at either end. The case is three inches long and holds enough powder to drive a 750-grain bullet — that's nearly two ounces of lead — from the muzzle at a speed of almost a half-mile per second. Even from a 15-pound rifle, the .577 is best fired while standing, just to keep your body flexible enough to absorb the recoil. Hemingway's buddy apparently didn't know that, and Hemingway didn't tell him. The man sat down, snugged the butt to his shoulder, squeezed the trigger and disintegrated his collarbone. Hemingway, never known for being abundantly sympathetic, found the whole thing hysterically funny. Abercrombie & Fitch did

Griffin & Howe has always been a shop for shooting men, from the obscure to such famous names as Ernest Hemingway, Gary Cooper, Robert Ruark, Dwight Eisenhower, Artie Shaw, Clark Gable, Elgin Gates, and legions of others.

not, and the shooting range was thereafter off-limits, even to their salesmen.

To devote maximum space to the retail business and also to shield the gunsmiths from well-meaning but distracting customers, Seymour Griffin, who remained manager and at least half-owner of the firm until he retired, moved the workshop to separate quarters. In the shop, the master 'smiths carried on their work and new generations of craftsmen learned the skills. Even into the 1950s and '60s, the masters mainly were Europeans, trained in the Old World tradition. Their influences remain. Josef Sovenyhazi designed a single-trigger mechanism that Griffin & Howe will still fit to any double gun. It's a simple, reliable device with a mechanical shift and an inertia block to prevent doubling. Josef Fugger, one of the country's finest artisans, was chief engraver. At a customer's request, he once inlaid a three-eighths-inch gold likeness of a canary on the top tang of an elephant rifle. The man wanted a memorial to his recently deceased pet. At Griffin & Howe, as at all good custom shops, he got what he asked for, flawlessly executed. Fugger also was a teacher and trained some of today's finest engravers, among them Joseph Bayer, Bob Swartley, and Winston Churchill.

Seymour Griffin retired about 1958, and full ownership of Griffin & Howe passed to Abercrombie & Fitch. Some good years still remained, but by the early 1970s, Abercrombie & Fitch's fortunes were seriously on the wane. The company went into receivership in 1975. Fastened as it was to a dying firm and in those years largely neglected by Abercrombie & Fitch, Griffin & Howe might well have withered altogether but for the efforts of Bill Ward and John Realmuto.

Realmuto joined the company in 1958 and spent a few years on the bench as a gunsmith and stock-maker. In 1961, he became assistant manager and then, a few months later, manager of Griffin & Howe.

Bill Ward, with a degree from the gunsmithing school at Trinidad, Colorado, came on board in 1968 as an apprentice 'smith. After five years at the bench making the various small metal parts that go into custom rifles, he worked as an assistant gun-buyer and in 1976, bought Griffin & Howe from the Abercrombie & Fitch receivers.

Reorganized under Bill Ward as president and John Realmuto as vice-president, the company moved out of the Abercrombie & Fitch building and into 589 Broadway, in the SoHo section of Manhattan. The shop, which had for several years been housed in the Abercrombie & Fitch warehouse in New Jersey, was installed in a connecting building that fronted on Mercer Street. In a sense, Griffin & Howe was back where it had been started — a privately-owned firm dedicated to the best standards of quality.

But it was a different world from the one Seymour Griffin and James Howe had faced 53 years earlier. The custom-rifle business had faded to an echo, and both present and future belonged to shotgunning. Bill Ward: "Griffin & Howe still has a reputation as a rifle shop, and, of course, we still build rifles. But that's a relatively small part of the business. The shotgun trade is by far the most active. Some of our customers hunt worldwide; a great many shoot at private hunting clubs. But what nearly all of our customers have in common is a desire for quality shooting and hunting. They're interested in refining their shotgunning, using a good gun that's well fitted, and properly set up. That's what we try to provide."

To that end, Griffin & Howe offers everything a shooter could want. You can buy a gun, have one repaired, or custom-stocked to your own measurements. They sell cases of various kinds, cartridge bags, handguards, and screwdrivers (or turnscrews, as the English call them) precisely ground to fit gun screws. As of this summer, you'll be able to buy American-made 2½-inch shotshells under the Griffin & Howe trade name.

They can even help you fine-tune your shooting. Twice a year, during the first two full weeks of May and September, the company sponsors a shooting school in the New York metro area. There you can learn from Rex Gage, formerly head shooting instructor at Holland & Holland, or from Ken Davies, Holland's current professor of shotgunning. A Griffin & Howe team will even come to your local gun club for a three-day shooting seminar. Any member of the staff can give you the details.

The small suite of showrooms on West 44th Street, where Griffin & Howe moved early last winter, is clearly a shotgunner's world. Row upon row of fine English guns, with a few from America and the Continent mixed in, stand in cherrywood cabinets. On one wall, a set of ceiling-high shelves holds more, all in fitted leather cases. In another room, ranks of rifles face yet more shotguns across a narrow aisle. Some of the rifles are English doubles, others trim bolt-actions. A few of these, so distinctive

Bill Ward, president and owner, in one of the new Griffin & Howe showrooms on West 44th Street.

that you don't even need to look for it, will have Griffin & Howe engraved on the barrels.

After the indifferent jostling you often get on a New York City sidewalk, the atmosphere is a particular treat. It's quiet, for one thing, and the staff treats you with cordial respect. Any of the salesmen — David Cosby, John Pfeiffer, or Rod Hatch — will show you a $25,000 gun or a $20 pair of snap caps and never make you feel as if they have something better to do. A lot of backwoods shops that survive on used pump guns and .22 cartridges could take a good lesson from that. If you're interested in having a gun restocked, John Realmuto will give you a fitting on one of Griffin & Howe's two Boss side-lever try-guns. Bill Ward is likely to be there, too, either at his desk in a corner of one of the showrooms or talking with a customer. They all are shooting men, and they know their stuff.

When I visited Griffin & Howe last December, the shop was still in SoHo, but a move to Bernardsville, New Jersey, was on tap for the spring of '87. No doubt the ambiance will be the same, a Dickensian clutter that every first-rate workshop ought to have. Riding up in a clattery freight elevator, I imagined a scene out of the old "Skilled Hands" Parker brochure — grandfatherly types with wire-rimmed spectacles and wispy white hair, patiently filing actions and fitting barrels and stocks at old, oil-soaked benches.

What I saw actually wasn't much different. The benches were scarred and splotched with gun oil. Near the door, an old Abercrombie & Fitch double was clamped in a stock-bending jig, its wrist steaming over a hot-plate and a coffee-can full of bubbling water. A matched pair of Purdey .410s, graceful as Clemens Hornn swords, lay on the next bench, their newly-fitted stocks almost ready for the first coat of finish. Joe Bayer sat at his bench across the narrow room, cutting delicate scroll into a rifle's floorplate. The whole place smelled of walnut dust and linseed oil.

But of the six gunsmiths currently on the Griffin & Howe staff, only three are older men. The rest are young, a couple of them mere kids from my perspective. Still, young hands or old, the final proof is in the results, and in that shop the results are obvious. One of the young 'smiths showed me his current project, fitting a quarter rib to a Griffin & Howe rifle built on a Kimber action; from any angle, the rib, barrel, and receiver looked as if they'd been milled from a single piece of steel. First-rate craftsmanship is hard to find in this frantic world and isn't likely to be more abundant in the future. At Griffin & Howe, though, fine quality is the standard and shows every promise of remaining that way.

So, from this crossing-point where one road stretches to the past, another points into the future. It's been 64 years now since Seymour Griffin and James Howe got the whole thing started, a goodly chunk of time for any small business to exist. But time itself is meaningless unless it's given form and substance, and then time becomes tradition.

"We have quite a few second and third generation customers," Bill Ward told me, "and we try to give them at least the same quality of service their fathers and grandfathers got. But the best part, the most gratifying legacy of the Griffin & Howe reputation, is the man who comes in and says, 'I've heard of Griffin & Howe all my life, and I trust you to give me the best advice.' We intend to see that that man's children and even his grandchildren feel confident in saying the same thing."

That kind of tradition, like old money, speaks with a voice all its own. ♦

BERETTA
Five Centuries of Gunmaking

The first were 15th century Venetian arquebusiers. Three hundred years later Napoleon's army used Beretta arms to conquer Europe. Today, almost 500 years after it began, the Beretta dynasty lives on.

MICHAEL McINTOSH

Trompia Valley is a narrow groove in the foothills of northeastern Italy, hewn by the Mella River. Through much of its length, which isn't far, there is scarcely a square yard of naturally level ground. It's a difficult place to grow crops, and in any case, the growing season is short. Unlike other valleys in that part of the world, Val Trompia doesn't even grow first-rate grapevines. But inhospitable as it might be to agriculture or the luxury of a good local wine, Trompia Valley is a piece of geography upon which nature bestowed a particular gift.

The valley slopes are laced with rich veins of ore that yield nearly pure siderite, iron carbonate. It has some phosphorus and a healthy dose of manganese, and produces a tough, lightweight iron that's easily worked. In the days before steel, such amiable metal was immensely useful, and ore was being dug and smelted in the valley as early as the Middle Ages, perhaps even before.

With all that going for it, Trompia Valley can lay a fair claim to being the oldest gun-making district in the world. It's a claim that Suhl, Antwerp, Augsburg, and Nuremburg might well dispute, but there is no question that Val Trompia is the oldest arms-producing district to remain an important center even today. Brescia, which lies at the southern end of the valley, is home to some of the finest gun-makers in Italy, and Gardone, a few miles north, has Beretta, the oldest and most famous of all.

By the middle of the 16th century, an arms industry already thrived in Trompia Valley, supplying arquebuses, muskets, pike heads, breastplates, and myriad other items of military hardware to the courts of Europe. Gun barrels were a particular specialty among Val Trompian craftsmen, but the area was unique in that it could produce guns with a greater degree of self-sufficiency than anywhere else in the world. Ore and wood could be transformed into a finished gun without ever leaving the valley and without any component being imported.

As evidence of just how pervasive was Val Trompia's influence upon arms-making,

---

Tuscan fowling piece, circa 1725, with barrel signed "Giovan Beretta;" and Beretta over-and-under made in 1980.

national museums all over Europe and Asia contain guns locally manufactured but fitted with barrels signed by the *maestri da canne* of Gardone. Bartolomeo Beretta was one of them, and he represents the beginning of a family of gun-makers that would remain active in the industry for an astonishingly long time.

Putting a date to the exact origin of the Beretta company probably never will be possible. There is suggestive evidence leading as far back as 1450. Bartolomeo Beretta was a working barrel-maker in the 1530s. His second son, Giovannino, was by 1577 a master maker with his own shop. Because both tradition and law held that only the son of a master could himself become a master, the current company considers 1530 its founding date. A few years more or less scarcely matter; what's most intriguing about Beretta is that it has survived as a family-owned business through at least 12 generations and is today one of the three or four preeminent gun-makers in the world.

Beretta's unbroken line of family gunmakers may constitute the oldest industrial enterprise in the world.

Inasmuch as Europe was the scene of countless wars and endless political uproar over those 400-odd years, it's hardly surprising that Beretta has from the beginning built military arms. But through all that time, Beretta has produced sporting arms as well. From the 15th through the 18th centuries, the superb Gardonese barrels, lightweight and marvelously strong, were the heart of the finest sporting pieces in the world. A fair number of those bore the Beretta name.

Almost without exception, Gardonese barrels were smooth-bored. Beretta, in fact, made no rifled barrels until the beginning of World War I — which meant that military and sporting barrels could be made interchangeably, with no alteration of technique and without interrupting production. Since everything was built by hand, the shop could fill military orders and switch to sporting guns without stopping for breath.

Beretta's guns followed the same evolutionary stages as all the rest. There were matchlocks, wheellocks, and flint guns. Naturally, all were muzzle-loaders, although like others, the Berettas experimented with breechloading designs. Giovanni Antonio, the third generation of gun-making Berettas, in 1641 presented the Venetian government with a breechloading cannon of his own invention. After testing it, the government awarded Beretta 200 ducats, a 20-year annuity of ten ducats a month, and promptly forgot about the whole matter.

The Venetian Republic, which governed nearly the whole of northern Italy from 1426 until the rise of Napoleon, took a protectionist view of its industry, certainly including the arms industry in the Trompia Valley, and even though the valley itself was periodically wracked with internecine social struggle, the national economy remained stable. The gun-makers freely carried on their work, their rights of trade supported by the government. After 1797, when Napoleon annexed Lombardy and dismantled the old republic, Val Trompia was governed by the French, and the arms industry turned virtually all its resources to supplying weapons for Bonaparte's Armies of the Eagles.

Eighteen years later, Napoleon went down in final defeat, and Lombardy became part of the Austrian Empire. The market for military arms stopped instantly. For the gun-makers in

After centuries of producing military weapons, Beretta knows how to make rugged and reliable centerfire rifles. Top: Beretta's 502 DeLuxe bolt-action and SSO Express Rifle.

the Trompia Valley, the world was a new and somewhat hostile place. The obvious course was to renew the civilian markets, but those were in sad disrepair. The vast Oriental market, to which the Val Trompia traditionally had supplied gun barrels, was gone altogether. Most of the others had been usurped by the French, English, and Belgians. To make matters worse, the European economy was sluggish, and the Austrian government was hungry for taxes.

Beretta's position was no better than anyone else's. The family still had its workshops, but without some means of competing on a broad scale, its prospects were dim. But as sometimes happens, the right man was in the right place at the right time.

Pietro Antonio Beretta was born June 18, 1791, heir to eight generations of gun-making tradition. Like his forebears, he had all the skills of a master barrel-maker, but he also owned an astute sense of how to meet the demands of a largely uncertain future. Almost before the cannon smoke had cleared from the battlefields at Waterloo, Pietro Beretta left the Trompia Valley to gather up the threads of a tattered market.

Beretta's huge plant varies from space-age automation to meticulous personal inspection of gunstock blanks.

He met with wholesale dealers, retailers, and importers all over Italy. Inevitably, there soon was some demand for martial arms from the Hapsburgs in Austria and their allies, but it was scarcely a trickle compared with the old days of the Venetian Republic. No doubt Pietro Beretta was pleased that his workshops were turning out barrels of any kind, but he was shrewd enough to know that the Val Trompian industry and the Berettas in particular could hardly hope to prosper without a strong showing in sporting guns. His pursuit of that market, won largely through the network of contacts he forged over the next few years, would in time form the foundation upon which the company still stands.

Even the name he gave it in 1832 remains today — *Fabricca d'Armi Pietro Beretta*.

He would not live to see it come to full flower, but Pietro Beretta virtually assured the firm's future. In 1850 he bought back a smithy that his father had sold in 1814. The two buildings, with their forges and workshops, were important enough, but along with them came the rights of use for such canals and watercourses as might provide both transportation and a source of power. The full significance of the purchase would not become apparent for more than a generation.

Pietro Beretta died in 1853, and in due course, control of the company devolved upon his son Giuseppe. Here, too, was a man who knew that merely manufacturing barrels or locks was no assurance of survival in a world grown increasingly complex. He strengthened and expanded the marketing network his father had put together and transformed Armi Beretta from what still was basically a barrel shop into a full-fledged manufactory.

The gun-making industry in the Trompia Valley had always been largely cooperative. The Gardonese masters made barrels; craftsmen in other villages made locks, springs, mountings, screws, and other metal parts; the majority of guns were assembled, stocked and finished in Brescia. By about 1870, Beretta had consolidated all phases of gun manufacture into one operation. With that, Armi Beretta became unique. In an 1878 letter to *La Borsa,* a newspaper published in Naples, Giuseppe Beretta claimed that his firm "alone possesses the remarkable quality of total production: that is, it brings to its premises raw iron and wood, and sends out finished guns ..."

From 1850 to 1860, Beretta produced about 300 sporting guns each year; by 1881 annual production was 8,000 shotguns. Since then, the firm has sold in excess of one million over-and-under guns.

The majority of them, by about 1880, were sporting guns, although Beretta continued to pursue military contracts. A few simple statistics clearly show where Giuseppe Beretta was headed. From 1850 to 1860, Beretta produced about 300 sporting guns each year; by 1881, annual production was 8,000 finished shotguns and nearly 2,000 double-barreled pistols. Moreover, the firm annually sold about 2,500 pairs of barrels and accompanying parts to other gun-makers. There were 200 employees in the factory that year, and the production rate was high enough that a good deal of work was

contracted out to other firms around Gardone.

Beginning about 1873, primary emphasis went to breechloading shotguns, and before the decade was over, they were being sold in every corner of Italy, in Greece, Turkey, Tunisia, and Egypt. The 1887 catalogue lists more than 100 different shotguns. About 70 are breechloaders of either centerfire or pinfire type; some 50 of these were of Beretta's own make, the others Belgian and English imports. There are sidelocks of various designs, with bolting systems ranging from top-lever snap actions to underlevers with wedge bolts.

The first hammerless guns appeared in the catalogue of 1893, and these included boxlocks as well as sidelocks. The vast variety of Beretta sporting guns would continue to flourish until the beginning of World War I, when the firm made its first serious — and as it turned out, prophetic — foray into the design and manufacture of automatic weapons. The results of that make a story in itself, for Beretta continues to occupy an eminent place in the international military-arms industry.

Given Italy's involvement in the Great War, the demand for sporting firearms naturally dwindled after 1914, only to revive shortly after the Treaty of Versailles. By then, Armi Beretta was in new hands.

Giuseppe Beretta died in June 1903, and leadership passed to his son, the second Pietro. In less than 30 years, building upon what his grandfather and father had begun, Pietro Beretta brought the firm into international status. To ensure that production could keep pace with the world markets he so successfully curried, he expanded the factory, installed new machinery and, taking advantage of water rights the first Pietro had purchased long before, built two generating plants on the Mella River to secure his own source of electric power.

The proliferation of sporting-gun designs

Pietro Beretta at the wheel of his DeDion-Bouton in 1906 with target-shooting companions.

abated somewhat between the two world wars, but there was variety enough to reach almost every niche in the market. Breechloaders naturally predominated, but muzzle-loading percussion guns remained in production until 1923. The 1938 catalogue even featured a special insert offering pinfire guns, ranging in finish from plain to lavishly ornate. As early as the mid-19th century and probably even before, Beretta had built a number of over/under muzzle-loaders. No doubt well aware of John Browning's Superposed and probably sensing what an impact the over/under gun would have in the future, Pietro Beretta introduced his own version in 1932, called the Model SO. The name is still used for Beretta's best-quality sidelock over/unders.

Even with so great a diversity of types, virtually all of the Beretta double guns built in the 20th century share a common basic design. The monobloc system first appeared about 1903 and has been used ever since. The breeches of both barrels, the barrel lump, and rib extension are milled integrally from a single block of steel. The breeches are countered-bored to accept the barrels. The block is heated to about 350 degrees Centigrade, and the barrels, their chamber-ends lathe-turned to form sleeves, are inserted. As the block cools, the shrinkage, combined with a special-alloy solder, bonds barrels and breech as firmly as if they'd been machined all of a piece.

Of itself, the monobloc system doesn't necessarily produce a stronger gun than one built the more traditional way, but it does offer some practical manufacturing advantages. If a break-action gun is to open, close, and lock up properly — and continue to do so for a lifetime — it's critical that the breech, barrel lump, and bolting notches all fit together in precisely the right way. Assembling them as separate parts is a demanding job; if they're all made in one piece to begin with, the task is simpler, requires less handwork for fitting, and essentially amounts to higher quality at lower cost.

The monobloc system isn't the only way to build a high-quality gun, but it's an excellent way to build one that doesn't cost your left arm and your firstborn.

Providing quality is something that Beretta does very well indeed. Any Beretta gun — whether it's an assault rifle, a pistol, a pump or autoloading shotgun, or a meticulously finished, best-quality double — is an admirable piece of work. The difference between a custom-built game gun and a production side-by-side or over/under is largely a matter of cosmetics and wood. Everything else, the steels and the functional mechanics, is identical. To my mind, that makes the lower-grade Berettas among the best buys on the world market.

And it's the world market that really tells the tale of Beretta's success. American sales of sporting guns are respectable but not remarkable; it's just the reverse nearly everywhere else. Over the past 20 years or so, the Italians have all but taken over the market for top-quality target guns, and Beretta is preeminent in almost every country except America, claiming a nearly 70-percent share of the European market. You won't see all that many Berettas on the firing lines at the big-time American tournaments, but don't let that fool you. The rest of the world shoots more Berettas than anything else.

The reason is simple: Beretta builds a superbly rugged, thoroughly reliable gun. That comes from a deep understanding of what a shotgun ought to be, a commitment to building it that way, and from having been around long enough to figure out how to do it. ♦

Spanish
TREASURE

In the foothills of the picturesque Pyrenees are the last of Spain's handmade-to-order gunmakers — producers of fine side-by-side shotguns that don't cost a fortune.
TERRY WIELAND
PHOTOGRAPHS BY J.M. ZABALA

At 81, Pedro Arrizabalaga is the dean of one of the most exclusive clubs in the world: the fine custom gunmakers of Spain. Now living in retirement, Senor Arrizabalaga occupies his time playing guitar and a regular game of billiards. But every so often, he places his distinctive Basque beret above his firm Basque features, and trudges from his apartment up through the narrow, winding, climbing streets of Eibar to a nondescript doorway, then up two flights of stairs, and into the small, dusty, cluttered gunshop with the tiny sign on the door that reads simply: Pedro Arrizabalaga.

Walking directly into the office of the company's manager, Jose Garate, the retired master will open the gun cabinet, take one of the finished shotguns, and begin a slow, careful, ruthless examination.

"He began this when I first started with the company as an engraver," Garate says. "No gun left the shop without his personal approval. If he found a fault, he handed the gun to me without a word, to see if I could find it, too.

"If I found the problem, he would say 'You're learning;' if I couldn't find it, he'd say 'You're young yet — you'll learn.' "

This was the beginning of Garate's long initiation into the exclusive club that now includes only Arrizabalaga, Armas Garbi, AyA-Diarm, Arrieta, and Grulla Armas. These are the last to carry on a Spanish gunmaking tradition going back 500 years — handmade-to-order gunmakers, producers of fine side-by-side, sidelock double shotguns.

Sitting in the middle of a long table in the trap-range clubhouse, surrounded by the gunmaker-managers of the other four companies, in a sea of cigar smoke and talk of Spanish politics, no man is more respected than Pedro Arrizabalaga. No company is more respected as a gunmaker. It is not the largest, nor the most expensive. Some of the others may be as good. No one is better.

When we arrived that morning at the trap range at Arrate, perched on a mountaintop above Eibar, the mercurial Basque weather was changing from threatening rain to glistening sunshine, clouds sometimes engulfing the mountaintop, and other times blowing away to reveal the view, in every direction, 15 to 20 miles through the foothills of the Pyrenees. The trap range and its adjacent box pigeon range sit close beside the Shrine of the Virgin of Arrate, the favorite wedding site for Eibar's courting couples.

One by one, the cars arrived, casually dressed men unloading guncases and shipping boxes from their trunks, the guns carefully unpacked and laid out in a row behind the firing line.

Arrieta's Model 803, with its fine Purdey-style engraving, reflects the skilled craftsmanship of Basque gunmakers.

Senor Arrizabalaga walked slowly along, picking up the guns that bear his name, a ritual inspection. His eyes roved along the line, resting critically upon each gun in turn, but until invited, touching only his own. Jose Luis Usobiaga, chairman of Grulla Armas, broke open the package of ammunition. Jose Garate handed me an Arrizabalaga gun styled after the Purdey, right down to the engraving pattern and the slightly oversized splinter forend.

A sidelock, straight grip, double-triggered game gun is not ideal for trap. But so what?

"LISTO!"

"Listo!" the muffled reply.

"PLATO!"

The yellow clay pigeon sails out, climbing against the black background of a line of pine trees. BOOM! A clear miss, and the disk descends in a graceful arc down the mountainside.

Polite to a fault, we all insist that someone else be next. Manolo Santos, commerical director of Arrieta, picks up one of his company's live pigeon guns and calmly breaks the bird almost as soon as it leaves the trap. Walking back to the line, he hands the gun to Senor Arrizabalaga, inviting inspection.

Arrieta is the largest of the independent makers (AyA is now part of a larger company) with about 20 employees. It turns out anywhere from 500 to 1,000 guns a year, some in batch lots built for dealers. It has the most extensive range of models, and in many ways is the most progressive.

In business since 1940, it is now owned by the Arrieta brothers, Jose and Victor. While Santos runs the business end, Victor is production manager, and Jose is in charge of quality control. Their shop, in the small town of Elgoibar, just outside Eibar, is a model of old-world craftsmanship and efficiency. Classic music fills the large room. Every bench is clean and orderly. Off by himself, sunlight streaming in from bright windows on three sides and the traditional canary singing at his shoulder, Arrieta's engraver, Eduardo Aramburu, carefully inlays a 24-karat gold "2" onto the barrels of an almost-finished shotgun. The air smells of gun solvent and linseed oil.

As with most aspects of shotgun making, the genius of Spanish engraving lies not in a unique style or method, but flawless duplication of the best work of others. Mostly, the Spanish engrave in the English style, with the traditional Purdey scroll pattern most popular, but they also do Belgian and German-style deep engraving. The best Garbi, the Deluxe, has an elaborate pattern with a naked cherub, a dragon, and gargoyles on the ears of the standing breech. The engraving is lovely, if your taste runs that way.

Where the Spanish do excel is in gold inlays. This traditional craft, called "damasquina," is taught at the gunsmithing school in Eibar, and is used to inlay numbers, initials, gold borders, and even scrollwork. Of the five gunmakers, Arrieta is noted for its damascened receivers.

Arrieta makes 13 models, all sidelocks (only AyA makes a boxlock of any kind.) A non-selective single trigger is available, and is usually put

on the live pigeon guns, but double triggers, straight grips, and splinter forends are the rule. Arrieta's prices range from 179,000 pesetas to 1,183,000 pesetas ($1,400 to $9,200 U.S.).

Senor Arrizabalaga eyed the smooth black Arrieta pigeon gun as he took it from Santos' hands. Alone among the four companies, Arrieta offers such innovative, non-English features as a black finish with gold borders, or even an unengraved coin finish. What the old man thinks of the gun's modern styling is anyone's guess, but no one could fault the workmanship, and he hands it back with a smile.

Jose Luis Usobiaga presses a Grulla into my hands, and I step forward once again.

"PLATO." Boom! The bird shatters. Applause. A modest smile and a silent "Thank God!" Jose Luis claps me on the back as I return the gun.

Grulla Armas (pronounced GROO-yuh, Spanish for crane), formerly Union Armera, is, along with Garbi and AyA, the best known among American gunners. Until recently, Grulla supplied The Orvis Company; Orvis is now supplied by Arrieta.

Still, the questions beg to be answered: Why are the best Spanish guns so little known in the U.S.? Why do they sell so few? In the case of the five, the answer lies in the fact that their custom-made, low volume, relatively high-priced product does not lend itself to the traditional manufacturer-importer-dealer relationship, and so importers are generally not interested. So, the makers depend on word of mouth and the knowledge and initiative of prospective buyers. Arrieta and Grulla both make a point of attending major shows in the U.S. to take orders; otherwise, their presence in the American market is sporadic.

The day we met, Jose Luis was at pains to explain some of the problems that his and other companies have in supplying shooters in the U.S., in the face of competition from bargain-basement imports produced by companies such as Rossi in Brazil.

"We are not automated, and we don't want to be," Jose Luis told me. "We make guns by hand, all sidelocks. Even our lowest priced gun is handmade, to order, to fit the customer. So how can we compete with Rossi and the others?"

How, indeed? One way Grulla competes with the other custom makers, if not with Rossi, is price. It offers 14 models, ranging upwards from 46,000 Pts. for the 209E. That's right, ladies and gentlemen: a handmade sidelock double with custom-fitted walnut stock, for about $400, FOB Eibar!

A Purdey it ain't, to be sure. But there it is. I haven't seen one in the flesh (being custom makers, none of the companies have many completed guns around, except those waiting for shipping), but Jose Luis explained how they are able to offer such a gun, and where money is saved.

The day I visited Diarm, they were finishing a matched set of six (six!) Seniors for a Spanish bank executive. That's $70,000 on the hoof.

"First of all, it is a monoblock construction, not chopper lump," he said. "That is the biggest saving. Its sidelocks are case hardened, with only a machine-engraved border. The stock is plain walnut (but walnut, nonetheless). And it is our only model where hand-detachable sidelocks and the H&H-type easy opening mechanism are not available."

Stockmaker Antonio Iriondo, 77, one of the Arrizabalaga's founders, continues to work on special projects at the Eibar shop.

"We can't afford to take the time to finish it as well as we might like, but otherwise, it is made by the same people who make our top guns."

Curiosity alone (for $400, how can you lose?) almost compels one to order one.

Grulla's higher grades compare, both in quality and price, to the models offered by the others.

Senor Arrizabalaga and Jose Arrieta broke away from the group at the firing line and wandered over to the edge of the box pigeon range. Far below, like a child's blocks that have rolled down a hillside into a jumbled heap in the valley, is the town of Eibar, gunmaking capital of Spain since the invention of gunpowder.

Eibar, the "villa armas," is to Spain what St. Etienne is to France, Leige is to Belgium, and Brescia to Italy.

It is probable that when Columbus discovered America, he was carrying weapons made in Eibar. They were making guns in Eibar when Wellington defeated the French, when the North fought the South, through two world wars and inumerable Spanish civil conflicts. Today, there are still dozens of manufacturers in and around Eibar, making shotguns, handguns, rifles, and military weapons.

The town even houses the Escuela de Armeria, a school originally devoted to the basic training of gunmakers who then left to take up apprenticeship in the scores of gunmaking companies of Eibar. The school has since branched out, and teaches everything from damasquina to high technology machine work with the latest automation equipment.

Surprisingly, given its devotion to gunmaking, Spain is not noted for innovation. It never produced a John Moses Browning or a Joseph Manton. Instead, Spanish gunmakers have taken the innovations of others and perfected their manufacture. Today, in the catalogues of the best makers, the terms "H&H-type" and "Purdey-style" abound. Their passion for English-style guns is not surprising. Historical ties between England and the Basque country go back centuries, the Basques trading their iron for British coal, supporting the firearms industries of both countries. Today, instead of steel, the Basques are exporting skills, and Spanish craftsmen labor in the workshops of more than one big-name English gunmaker.

Down through the ages, the names of famous Basque gunmakers reverberate: Aguirre, Aranzabal, Arizaga, Ugartechea, Sarasqueta, Zabala. In *The Shotgun Book,* Jack O'Connor spoke highly of Arizaga and AyA, of Ugartechea and Victor Sarasqueta. In a 1954 article on Spain, Charles Askins stated flatly "AyA is the best."

In recent years, the malady that struck American double makers in the 30s, the British in the 50s, and the Belgians in the 70s, has caught up with the Spanish: competition from low-priced guns, and the unwillingness of young people to go into the trade — at least, not as low-paid apprentices working long hours to learn the traditional skills.

Today, Ugartechea is a mass producer of lower-priced guns; Victor Sarasqueta is out of business. AyA survives as part of a merging of 20 firms (including Arizaga, although the Arizaga name has disappeared) into the new and modern gunmaking company, Diarm SA.

Pedro Arrizabalaga shares his six decades of gunmaking wisdom with, from left, Jose Alberto Garate, also of Arrizabalaga; Jesus Barrenechea, Armas Garbi; and Jose Luis Usobiaga, Grulla Armas.

In its ultramodern, 175-employee plant near the village of Itziar, overlooking the Bay of Biscay, Diarm manufactures an imaginative line of side-by-sides and over-and-unders, using techniques such as acid engraving to produce good looking, dependable, and low-priced guns that the Basque government hopes will ensure the survival of their traditional industry.

Of the companies making up Diarm, only the AyA marque will survive, and when existing stocks of AyA barrels and parts are used up, AyA will become a purely made-to-measure operation. The four-model line will have two distinctions: the only Spanish custom-made boxlock (at about $400) and the most expensive sidelock, the Senior, at $11,700!

The day I visited Diarm, they were finishing a matched set of six (six!) Seniors for a Spanish bank executive. That's $70,000, on the hoof. The beautiful walnut stocks were so closely matched they were practically indistinguishable, and the checkering and engraving were superb.

Interestingly enough, AyA is not offering a single trigger on any of its new models, regardless of price; the endless railing against Spanish selective single triggers that do not select seems to have had an impact. In fact, none of the five custom makers offer a selective single trigger. Rather than alter their traditional, proven product, the makers stick with double triggers, or at most, a non-selective trigger, which does not present the same difficulties.

I did not get a chance to fire an AyA; I made up for it with two Garbi guns. Garbi is the Holland and Holland of Spanish gunmaking. Even Pedro Arrizabalaga's eyes lit up when they alighted on the Garbi pigeon guns.

The company was founded in 1959 by five craftsmen; G-A-R-B-I is an acronym comprising each of their initials. The "B", Jesus Barrenechea, is now the general manager, running a shop of a dozen men, turning out four models, with one simple aim: "We want to be the best."

At least one person believes they've made it: King Juan Carlos I of Spain has stopped shooting his matched Purdeys and now uses Garbi guns exclusively (he has eight). What's more, his standard gift to visiting dignitaries is a custom-made Garbi shotgun. If you want one, and you are not a friend of King Juan Carlos, it will cost you between $1,500 and $10,900, with delivery in about three months.

In the past, Garbi has sold about 100 guns a year in the U.S., through the firm of William Larkin Moore in New York. Now they are finding, however, that more and more Americans are flying to Madrid, doing some shooting, and ordering fully custom-made guns through local gun dealers. It's cheaper, and it's a lot more fun.

The term "fully custom" is relative, however: if you have a yen for a Garbi with a mesquite stock, forget it. They only work in walnut. Like the other custom makers, their standard stock wood is Spanish walnut, a wood that is almost impossible to tell from French walnut, but is considerably cheaper.

"Sometimes a customer has his own stock blank that he wants us to use," Jesus said. "We've worked in American, and English, and even claro walnut. I don't believe any is as attractive as the Spanish, but it is a matter of taste."

The lowest-priced Garbi is the Model 100, which sells for about $1,500. The receiver has a classic, Purdey-style scroll pattern. The catch is, it is not hand engraved — the only Garbi which is allowed to go near a machine during manufacture (the *only* machine in the entire shop is a belt sander, used to remove the initial layers of wood from the stock). Model 100s are "sent out" to have their engraving done. Still, the engraving I saw was smooth and elegant. An expert would be able to tell the difference perhaps; I could not.

Garbi doesn't advertise the fact, but they will give their guns a case-hardened finish if the client prefers it; Jesus says most Americans do, while Europeans tend to like the bright finish. That, however, is about the only concession they make. The "walnut-only on a Garbi" attitude extends to the engraving on the receiver as well. If you want a plain coin or case-hardened receiver without engraving, the answer is no.

"We believe a gun of this quality must have engraving, and the engraving must be good," Jesus told me. "We will not put the Garbi name on a gun that does not meet our standards of what a fine shotgun should be."

That's class.

The firm of Pedro Arrizabalaga, however, has carried the whole concept of custom gunmaking the final step. If Garbi is akin to Holland and Holland, Arrizabalaga is the Purdey. They now make only one model: the best. Within limits of their own taste, they will make it as you want it. Jose Garate told me that a typical Arrizabalaga Purdey lookalike will cost in the neighborhood of $5,500 — about half the cost of a top-of-the-line AyA or Garbi.

The Arrizabalaga philosophy has evolved gradually. Ten years ago, with ten workers, the company produced about 250 guns a year; today, with 11 workers, they produce about 160 — the fewest of any of the custom makers. However, they also take the longest, with an average of eight months delivery, and most of their production is sold in Britain, France, and Spain. Except for a few American aficionados, who let only their close friends in on the secret, the Arrizabalaga is largely unknown in the U.S.

"We are producing fewer guns, not because of fewer orders, but because we are taking longer with each gun," Jose says. "Each gun must be perfect."

Not many gunmakers insist on showing you the innards of a shotgun as soon as you walk in; Arrizabalaga does. The workings of an Arrizabalaga sidelock are like a Swiss watch, smooth and polished and intricate. The metal rests heavy in your hand, and you don't want to put it down. Even through a magnifying glass, the working parts of the lock showed no tooling marks. None.

Arrizabalaga's clientele consists of obscure but fantastically wealthy European noblemen of ancient family, merchant bankers, the occasional American industralist. Surprisingly, they sell many guns in Britain, where their ability to

Located at the edge of the Pyrenees, Eibar has been the gunmaking capital of Spain ever since the invention of gunpowder.

make a Purdey lookalike at a quarter of the price is appreciated by those who don't have $20,000 and four years to spare.

While I was there, I met another of the founders, master gunmaker Antonio Iriondo, age 77. His son has taken over from him, but he still comes down on evenings to work on special projects for special clients. Grinning under his beret, Senor Iriondo proudly pulled out a prized momento: a postcard from Col. Charles Askins, in uniform, with rifle and a tiger hanging up. It reads: "Greetings from Indochina. Askins."

Senor Iriondo is reputed to do the finest oil finish of any Spanish stockmaker. It is a secret only his son knows, or will know. This tradition, however, as elsewhere, is in trouble.

"We are the last," Jose Garate told me. "Young people today are not interested. Look around. The men in all our shops are in their 40s, and there is no one coming up. When we go, I am afraid it will go with us."

Lunch at the Arrate shrine and trapshooting club is over. Lazily, full of serrano ham and white asparagus, fine red wine, and Duke of Alba brandy, with the smell of expensive cigars still hanging in the air, we rise to leave. For Senor Arrizabalaga, it has been a long day, and tomorrow he leaves for a month on the Costa del Sol.

Outside, the sky has cleared completely. In the brilliant sunshine, you can see for miles through the Pyrenees, and down through the centuries. This special event, organized for my benefit, is the first of its kind. We joke that we should make it a tradition. The Arrate and Eibar Gunmaking, Trapshooting and Luncheon Club (Membership Restricted).

I promise to return in 20 years to see how the Spanish gunmakers are faring, and Pedro Arrizabalaga says with a smile, "I'm sorry I won't be here to see you."

One by one, the cars pull away, descending in line along the road that clings like a ribbon to the mountainside, down past Basque farmhouses and Basque sheep that dot the vertical fields. A browsing goat eyes the cavalcade, and dances off through the pine trees.

Down into the valley of Eibar we go, one after the other. Arrizabalaga. Garbi. AyA. Grulla. Arrieta. The most exclusive club in Spain.♦

You can go to Eibar in person to buy a Spanish side-by-side, or you can order direct from the U.S. and have it sent. Either way, you pay the same price a dealer would. The prices quoted are FOB Eibar, with air freight, insurance, duty, and taxes extra. Even so, you save a bundle.

Fitting tends to be a trade secret with each company, and they would all prefer to have you come to them to be measured. If you cannot, they tell you what measurements to take, and calculate dimensions accordingly.

Language can be a problem. If you do not speak Spanish, an interpreter is inexpensive insurance. Interpreters and translation services are available from: T.I.S.A., San Vincente, 8, Edificio Albia II, 7th Floor, 48001 BILBAO, Spain.

For more information about the gunmakers and their products, write: Asociacion Armera, Plaza de Unzaga, 5, First Floor, 20600 EIBAR, Guipuzcoa, Spain; *Armas Garbi*, Urki, 12-14, EIBAR, Guipuzcoa, Spain; *Manufacturas Arrieta*, S.L., Barrio Uransandi, Apartado 93, ELGOIBAR, Guipuzcoa, Spain; *Pedro Arrizabalaga*, S.A., Errekatxu, 5, EIBAR, Guipuzcoa, Spain; AyA, *Diarm SA*, Poligono Industrial de Itziar, 20820 DEBA, Guipuzcoa, Spain; *Grulla Armas*, Apartado 453, Avenida Otaola, 12, 20600 EIBAR, Guipuzcoa, Spain.

The Hard Times WONDER

Introduced in 1932, the Remington Model 32 was enthusiastically greeted by sportsmen seeking a low-priced, well-made gun. But the Model 32 proved to be much more.

MICHAEL McINTOSH

Remington Model 32 Trap Special with 32-inch barrels. Photograph by Ingbet Gruttner.

When the Great War was over, America revelled like an amusement-park gone out of control. She glittered and danced in the arms of jazz and bathtub gin, lavished herself with all the baubles money could buy. For the American shotgun, they were golden years. Wealthy sportsmen demanded more and more elaborately finished guns and were willing to pay the prices till then unheard of. The Sousa Grade Ithaca sold for $700, the Deluxe Grade L. C. Smith for $1,000. In 1923, A. H. Fox advertised its GE Grade gun, built on special order, for $1,100. Parker topped them all with the fabulous, $1,500 Invincible.

The party lasted just over ten years, until October 1929, and then the pleasure palace turned from gay to grim. The gun makers were staggered by the Great Depression. Some sold out to larger, more stable companies. Others simply folded. High-grade guns for sale at a pittance gathered dust in pawnshops and gun-club racks.

But while the demand for highly decorated shotguns hit rock bottom in the early 1930s, the gun market itself was far from dead. Hunting and shooting were no less popular than before, and well-built guns that could be offered at reasonable prices continued to sell. Mechanical guns — pumps and autoloaders — had been growing steadily more popular since the turn of the century and by the late '20s clearly dominated the sporting market. But even in hard times, there was still a place for the double.

Target shooters, especially, preferred the double's superb handling qualities, which no repeater could match. Trap shooting and the new game of skeet, invented just after World War I, attracted legions of gunners. Shotguns tailored to the games opened a whole new market, and the gun makers were quick to seize it.

As he had often done during his long career, John Browning showed other makers the way. His Belgian-built Superposed, the last gun Browning designed, appeared on the American market in the late '20s and immediately caught the target-shooters' fancy. Among American makers, Remington Arms was the first to recognize the over/under as the coming thing and in May 1930 began designing the first machine-made over/under shotgun ever built in America. It would be named the Model 32.

The new gun was the work of Crawford C. Loomis. He had come to Remington in 1912, first refining the Browning-designed Model 11 autoloader and going on to become one of Remington's most prolific designers. As was the custom of the time, Loomis patented his designs in his own name; for the Model 32, he was granted a total of six, dating from March 3, 1931 to July 14, 1936.

When it appeared on the market in March 1932, the new Remington was greeted as the herald of a new age. F. C. Ness, writing in the May 1932 issue of *American Rifleman,* had this to say: "Its advent marks the beginning of a new era for shotgun enthusiasts because the low price achieved at last places the over-and-under type of gun within the reach of the average gunner."

As it turned out, he was right.

In that first year of its life, the Model 32 sold for $75, but it was more than just a well-made gun at a good price. Crawford Loomis's rich imagination and technical brilliance had come up with features never seen before on a shotgun of any kind.

The over/under gun poses some unique problems in designing a way to fasten the action. The rib extension and top hook that serve so splendidly on a side-by-side double are almost hopeless for an over/under. In order to accommodate the vertically stacked barrels, the over/under's frame has to be deep, and there simply isn't room to add a top hook without making it deeper still. Browning chose to bolt his Superposed with a shallow barrel lump and sliding underbolt; that he had to split both lump and bolt to make room for the cocking lever suggests the approach is a compromise. Crawford Loomis saw a better way.

His solution is ingeniously simple. Instead of using hooks or pins or crossbolts, Loomis designed a sliding cover that holds the barrels in place from the top. Operated by a conventional top latch, the cover rides in slots milled into the top of the frame and engages stout rails on the sides of the upper barrel. Bearing surfaces are tapered to compensate for wear. Because it is linked eccentrically to the top latch, the cover travels slightly less than a quarter-inch forward and back and not only locks the action but also shrouds the junction of barrels and breech, protecting the shooter from gas blowback in the event of a pierced primer or ruptured case head.

Instead of pivoting on a traditional hinge-pin, the Model 32's barrels ride on heavy, rounded studs on each side of the frame. Pivot points are semicircular notches on the sides of the lower barrel. This allows a shallower frame, for one thing, and, for another, takes optimum advantage of the over/under's inherent superiority in recoil dynamics.

Unlike the bores of a side-by-side double, which lie in a plane well above the shooter's shoulder, an over/under's barrels — the lower one, especially — align virtually straight with the heel of the buttstock. Moreover, because the barrels of all double guns must shoot to the same point of impact at a given distance, bore centers are farther apart at the breech than at the muzzles. Recoil force, therefore, tends to drive a side-by-side's buttstock upward while at the same time torquing right or left. An over/under, on the other hand, tends to recoil nearly straight back against the shoulder. The slight downward torque of the lower barrel — customarily the first fired — directs recoil toward the shoulder and away from the cheek.

These phenomena usually aren't of much consequence in the game fields, where shots were fewer, but recoil is something to be reckoned with on the firing line at trap or skeet, and that goes a long way toward explaining why the over/under is the choice of so many of America's serious target shooters.

If there's an equivalent in game shooting, it's probably driven birds, though the over/under never has been as popular in England as the tra-

Among American gun makers, Remington Arms was the first to recognize the over/under as the coming thing and in May 1930, began designing the first machine-made over/under shotgun ever built in America. Named the Model 32, its advent would mark the beginning of a new era for shotgun enthusiasts.

ditional side-by-side. Having come away from a hot corner in a dove field feeling as if I'd gone three rounds with Muhammad Ali, I can imagine what a treat it must be to shoot incoming pheasants at high angles all day. But at Sandringham one day in 1886, Lord de Grey fired his Purdey side-by-sides almost 1,300 times, played cards until five o'clock the next morning and remarked that he'd never felt better in his life. Among my hunting partners, it's common knowledge that Lord de Grey was a whole lot tougher than I am.

Another novel feature of the Model 32 is the absence of fillets between barrels. Where Browning soldered thin strips of steel from muzzles to breech, Crawford Loomis, evidently thinking of target shooters, left open space. The main reason is heat dissipation. Gun barrels heat rapidly in target shooting, which produces both visible heat waves and physical changes in the barrel itself. You can lose a target in the shimmering mirage of heat distortion. A hot, expanding barrel firmly fastened to a cooler one can cause a noticeable change in the shot swarm's point of impact.

The Model 32's separated barrels expose more surface to the air, promoting optimum heat loss by convection. Leaving off the fillets also reduces the weight of the barrels by an ounce or two, which permits slightly thicker-walled tubes for the same balance and overall weight. In the prototypes and perhaps in a few of the first production guns, the barrels were rigidly joined at the muzzles, which did nothing to solve the heat-expansion problem. This was remedied early on with a slip-ring barrel hanger, a principle the factory called the Remington Floating Barrel. The hanger is soldered to the upper barrel and simply forms a loop around the lower one, so that either barrel can expand and contract without affecting the other.

The earliest 32s had double triggers, but by 1937 all of the target guns were made with an excellent single selective trigger, factory adjusted for a 3½-to-5-pound weight of pull. In 1938, the single trigger became standard for all models. Unlike Browning's single trigger, with its recoil-set sear block, the Remington trigger is mechanical and shifts instantly from one sear to the other if the first hammer falls on a dud shell or an empty chamber. The barrel selector is a button in the forward end of the trigger. Lock time is extremely fast, and both sears break with a crisp, clean release that makes a lot of newer guns seem pretty shabby by comparison.

The Model 32 was built only in 12 gauge. All 32s were made with selective ejectors. The safety, a conventional thumb-button on the top tang, is adjustable. There are three tapped screwholes in the side of the top tang, accessible when the buttstock is removed. By moving a setscrew from one hole to another, the safety can be made automatic or manual or it can be locked out altogether. If you fail to shoot at a target in trap or skeet because you forgot to take the safety off, it's counted as a miss, and some trap guns are made with no safety at all. That's fine on the target range, and trap guns are dandy tools

All told, some 6,000 Model 32s were built — on the average about 600 guns per year. Such small production runs posed thorny problems with only expensive solutions.

for high-flying doves and ducks. But a field gun without a safety is a disaster looking for a place to happen. Some shooters prefer a safety that automatically clicks on each time the top latch is opened; others don't. With the Model 32, you can have it any way you want it.

Though meant to be a high-quality gun at a modest price, the Model 32 was built in several grades, some of them wonderfully ornate. The lowest was the 32A Standard — a plain gun with skillfully checkered, straight-grained American walnut stocks, decorated only with a bird dog roll-engraved on each side of the frame. The 32 Skeet, a 32S Trap Special and 32TC Target models were similar in appearance although stocked with progressively fancier walnut. The high grades — 32D Tournament, 32E Expert and 32F Premier — were treated with superb hand engraving, fancy checkering and highly-figured wood.

A, D, E and F grades were available with 26-, 28- or 30-inch barrels, either plain or with an optional solid rib. The skeet gun could be had with 26- or 28-inch tubes; plain barrels were standard, both solid and vent ribs available at extra charge. The TC Target came fitted with a vent rib on 30- or 32-inch barrels. Extra sets of barrels were available for all of them.

Although the Model 32 never reached the point of being absurdly expensive, rising manufacturing costs inevitably escalated the prices. The Standard Grade cost $99.50 by 1936. The 1939 catalogue lists the Standard at $126, the 32D at $276.50, the 32E at $326.50 and the 32F at $411.50. Custom stock dimensions cost an additional $15 for Standard, skeet, and TC guns, but were free for the high grades. Extra barrels cost $60 a set.

A year later, as war once again stalked the world, the Standard gun was up to $153.55,

A Model 32A from the 1937 Remington catalogue.

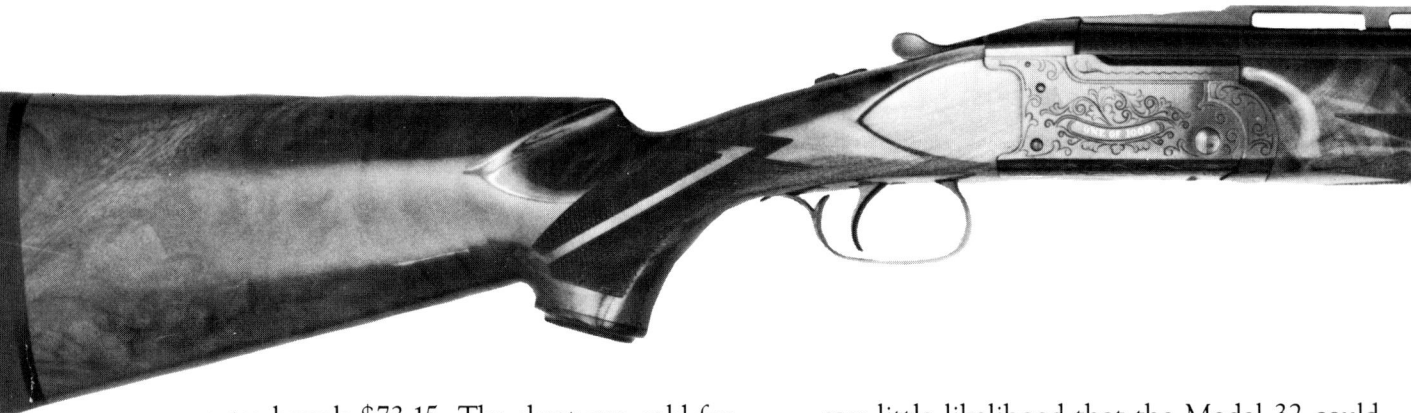

extra barrels $73.15. The skeet gun sold for $157.15 with plain barrels, $166.90 with solid rib and $182 with vent rib. The TC cost $188.70, the 32D $336.85, 32E $397.80 and 32F $501.35. The 1941 catalogue showed only the TC, at $164.10, and noted that a shortage of raw materials would delay delivery of all other models until the following year.

By then, though, it was a different world. In 1942, Remington converted its machinery to produce military hardware, and the Model 32 was discontinued. It appeared again in the 1947 catalogue but with an overprint that read "not available."

All told, some 6,000 Model 32s were built — on the average, about 600 guns per year. Such small production runs posed thorny problems with only expensive solutions. Like all double guns, the 32 required a great deal of hand fitting, and labor costs were rising. Faced with that and with the options of either retooling a lot of machines to produce only a few guns each year or leaving them tooled up and idle most of the time, Remington cost accountants saw little likelihood that the Model 32 could compete in the post-war market.

In the late 1940s, Remington sold manufacturing rights for the Model 32 to the old German firm of Heinrich Krieghoff, which by then had moved from its home in Suhl to the newly created West German Federal Republic. In its new factory at Ulm, Krieghoff revised some aspects of Crawford Loomis's original design and began producing the Krieghoff Model 32 — such a close copy of the Remington that components of the top latch, the ejector system and the bolting mechanism are interchangeable.

Shooters embraced the Krieghoff as readily as they had the old Remington, and for nearly 30 years it was the world standard by which target guns were judged. In the late '70s, no doubt responding to competition from Italy, Krieghoff redesigned the gun again to create the K-80 series, now marketed as the Shotguns of Ulm. In profile, it's still the old Model 32.

Almost from the moment the Model 32 was discontinued, American shooters beseiged Remington with the question: Could the old gun

Remington introduced its Model 3200 in 1973. Though quite different mechanically, it has all the features that made the Model 32 great.

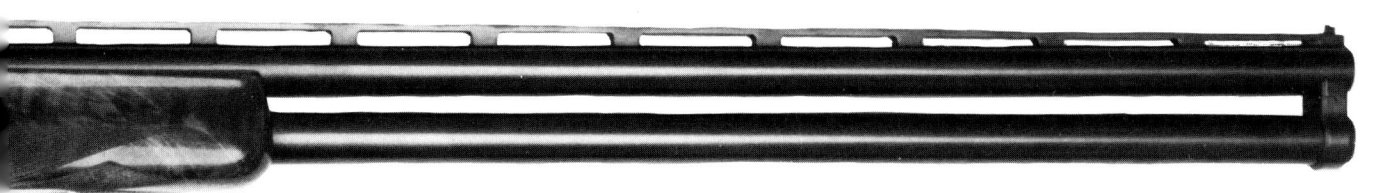

be made again? For years, the answer was the same: too expensive. But by the late '60s computer technology opened a whole new world of possibilities in gun design and manufacture, and a team of Remington engineers began working with designer John Linde, seeking to do what Crawford Loomis had done a generation earlier — design a well-made over/under that combined high quality and a reasonable price. The result, which appeared in 1973, was the Remington 3200. Though quite different mechanically, the 3200 has all the features that made the 32 great.

Not many Model 32s show up on the market these days. You can search for years without finding one of the high grades; not even the factory knows how many D, E and F grades were built, but there couldn't have been more than a handful. Presentation or custom-built guns seem to have been equally few, although there was at least one — a Model 32 with two sets of vent-rib barrels, a single trigger and fancy wood which the factory notes was "shipped to Kerr Sport Shop intended for Clark Gable." (Gable liked shooting and hunting almost as much as he appreciated the company of lovely women; those two interests came together in June 1936, when Carole Lombard gave him a DHE Parker 28-gauge as a token of her affection. It should happen to me.)

A Model 32 in original condition can be almost as hard to come by as a high grade. The wide, flat expanses of steel in the frame made it a favorite candidate for custom engraving; some of those guns, especially the ones engraved and inlaid in Germany, are magnificent, but upgrades aren't the real, factory quill. The majority of 32s probably were target guns, and target shooters being what we are, most of them seem to have been restocked, reribbed, reblued or otherwise altered.

But no matter. A 32 is a 32, and that's enough to fetch a handsome price. In the early '30s, a Remington advertisement showed a Model 32 and the question: "Would you pay $1,000 for this gun?" At the time, it was an effective approach. Nowadays, though, $1,000 is a bargain for just about any Model 32 you can find — if you can find one at all.◆

MY PURDEY
Much More Than a Gun

Endowed with seven decades of shooting tradition and the souls of two men, Purdey No. 21390 came as a gift and will be passed on as such. But for now, it is a living extension of its owner's nerves and brain.

GEORGE BIRD EVANS

If a man lives in the gun he loves and shoots, then my Purdey is endowed with the souls of two men, with a place reserved for me as Number Three. Those others were the gunners who shot it before me, this lovely thing alive between my hands.

This is the story of 12-bore Purdey No. 21390 with two pairs of 26-inch barrels, standard rose engraving, and third grip and clips to action. It has been mine to shoot for 25 seasons; I sometimes feel it has been mine for the 71 years since it was built, showing where hands have brightened the temper color on the sidelocks.

The gun was made for the post-Edwardian Philadelphia sportsman Lynford Biddle in 1915, the second year Britain was fighting the Hun. London gunmakers had converted to production of wartime matériel for the Crown, and Purdey's skeleton crew of their top gunsmiths built a minimal number of sporting guns for the duration. A thousand Purdey game guns and rifles had been produced in the three years prior to the outbreak of World War I; seven years passed before the next thousand were built.

Lynford Biddle fired thousands of rounds through No. 21390 for 26 years at upland game, wildfowl, and live pigeons. At his death he left it to his close friend and shooting companion Dr. Charles Camblos Norris of Chestnut Hill and later Fairhill, Pennsylvania.

It was characteristic of Dr. Norris to inquire in detail as to the specifications of the gun. In the flat oak and leather case, I found the following letter on Purdey's stationery, dated 19 October 1943:

We find from our records that the gun No. 21390 was originally chambered for 2 5/8" cartridges and was fitted with an extra pair of barrels. The No. 1 pair was bored right barrel choke, and the left full choke, whilst the No. 2 pair was bored right barrel tight cylinder, and the left modified choke.

The No. 1 pair shot with cartridges loaded with 33 grains of E.C. powder, with one ounce of No. 6 shot averaged 196 with the right barrel, and 207 with the left barrel in a thirty-inch circle at 40 yards. The No. 2 pair of barrels shot with cartridges loaded with the same charge and shot at same distance, averaged 147 with the right barrel and 194 with the left.

Trusting this gun will give you every satisfaction,

The letter was signed by Capt. James A. Purdey, then a director of the firm. On a lower corner of the letter Dr. Norris had written: *No. 1 right 72%, left 76%; No. 2 right 54%, left 71%. One ounce English No. 6 contains 270 pellets.*

I find Capt. Purdey's "tight cylinder" for 54% and "modified choke" for 71% interesting.

In 1956, on one of our shooting visits to Fairhill, I had the opportunity to examine "the little Purdey," together with Dr. Norris's other Purdey and his Churchill.

Having this gun in my gloved hands was like holding the Holy Grail. I will never forget the dim light in that bedroom in that big old house, with Charm the fat setter lying on a chaise lounge specially hers, Nellie the equally fat pointer lying on Dr. Norris's bed with her head on his pillow, snoring, and Dr. Norris standing at my side wreathed in blue pipe smoke, watching me thoughtfully.

For those of us who grew up with American shotguns, the feel of a London Best is a new experience with the nice balance between the gracefully slender straight "hand" of the stock and the small splinter fore-end. Soft light from the bedside lamp illuminated the fine English scroll engraving on the sidelocks, and the stock and action and barrels pointed like my extended arm. I can still see Dr. Norris's unspoken

George Bird Evans pauses on a morning grouse hunt with Belton, seventh generation of the Old Hemlock setters, and his third generation Purdey.

approval when I closed the hermetically tight breech and let the top lever down carefully with my gloved thumb.

Five short years later I was back in that gray stone house so like a country place in Norfolk, empty now without him and Nellie and Charm. Kay and I came away with the little Purdey, and my mind kept reverting to lines from a poem by Edna M. Cass in the *Shooting Times*:

Grandfather's left me his Purdey,
The lad said, half awestruck, half sad, ...

and the last line:

Grandfather left me much more than a gun.

In spite of the 30-year spread in our ages, Dr. Norris and I had much in common. As he had done in 1943, I wrote to Purdey's in 1961 and again in 1966. The reply to my second letter was written by Harry Lawrence, managing director:

I have checked on your gun No. 21390, which was built in 1915 and weighs 6 lbs. 2 ozs.

As you rightly say we build only one quality so therefore, there is no need for us to put a grade name such as "Royal" or "Crown" etc. Our gun in its standard state is finished with our fine rose pattern engraving but, of course, we do finish some of our guns with special engraving such as large scroll, game scenes, or deeply carved with gold inlays, etc., for which we naturally charge extra, but the actual gun itself is of one standard quality.

In the latter part of the last century we did for a short time build two or three qualities which were marked "B," "C," and "D" quality; very few were made for it was found to be a bad policy and was stopped.

Regarding the marks on the lower surface of the left barrel, the W.H. were the initials of the barrel filer, William Hill, for it has always been a policy of the Company that each individual craftsman should have the right to put his initials on the part of the gun he made, and this is still carried out today. You should, if you examine your gun carefully, find the initials of the man who made the action, the ejector etc., but you would have to strip the gun to find some of these...

The numbers 50047-48-49-50 are the actual numbers of each individual tube, for it enables us to check each tube back to any particular batch of steel and forging, for we keep perfect records of every tube used, and for which gun.

I have made some enquiries regarding the poem by Alaric Alexander Watts, but so far have not been able to find a copy, but I will continue with the search and should I find one I will immediately let you know.

The 6 lb. 2 oz. weight surprised me, for our local butcher had weighed the gun for me at 6 lbs. 7 ozs. It appears we may have been buying that butcher's thumb for years!

Some people have been over-awed when entering Purdey's Audley House in London's West End, but there is no place where lovers of fine guns receive more considerate attention, as with Harry Lawrence's personal effort to locate a copy of Watts' poem *Shooting Flying* for me.

Quoting the price of Purdey guns is a game

W.H. are the initials of William Hill, the barrel filer. The numbers 50049 and 50050 are for the individual tubes. S on early guns signified steel barrels vs. Damascus.

of keeping abreast of the exchange rate of currencies. When it was built, No. 21390 with its extras sold for £118.13.0; today an equivalent gun costs £21,300, or $32,802 exclusive of tax and insurance charges. What has not changed in a Purdey is quality, which like old money is unobtrusive but present.

A wood engraving of an 1886 hammerless Purdey game gun shows the same standard rose engraving that is on my 1915 Purdey and the same as on standard Purdey shotguns today, as nearly identical as hand work can be. There is a photo on page 31 of the March/April issue of Sporting Classics of a Purdey double rifle that bears the same rose engraving on its sidelocks, which have the same screwhead, pins, and discharge indicator as on my game gun.

The legend on the smooth top rib of my gun barrels states: *Made of Sir Joseph Whitworth's Fluid-Pressed Steel*, the steel that replaced the figured Damascus barrels in 1880s. The Purdey bolt and top lever locking mechanism patented in 1870, and the Beesley self-opening action patented in 1880 are used on Purdey guns today. Retaining something that was good over 100 years ago because it is still the best, is an immutable essence of tradition. The action on my Purdey is so fine-tuned it gives off a tone like a C-sharp harp string when rapped on the sidelocks.

Some collectors would, I suspect, undergo corrective surgery on their neck and shoulder rather than have a Purdey altered to their dimensions. But a gun, no matter how fine, serves its destiny only if it is shot and if it fits its gunner.

Most London Bests throw their patterns high at 40 yards, achieved with an "elevating rib," conceived by Joseph Manton. Unlike the elevated rib, which is raised the entire length of the barrels, the elevating rib is set high at the breech and slants to a small bead below the tops of the muzzles, causing an ascending or "elevating" line of fire in relation to the shooter's line of sight. To lower the muzzles to throw parallel with my vision, I had my gunsmith install a small platform with a white bead raised to my calculations, centering the patterns where I look.

Certain characteristics contribute to the gunner's performance — length of barrels, weight of trigger pull; balance, and gun weight — but gun fit can be said to be in the gunstock. When the Purdey came to me the stock required a bit more drop at comb and heel, considerably more length, and more cast-off. My gunsmith would normally have obtained the cast-off with hot oil and pressure, but hesitated to risk that because of the age of the wood, fearing it might not withstand the stress.

The blank for my stock had been cut from the root of a walnut tree that began to grow in France around 1700. The rough blanks are seasoned for six years before they come to Purdey's where they are kept for five or more years in the factory. To select outstanding patterns in the wood, the blanks are given a coat of stock oil, which brings out the configuration in the walnut.

I first obtained the extra length with a recoil pad but, dissatisfied with that, I sent the stock to a former stocker with Purdey's now in the States, who added an extension of French walnut and matched the grain with uncanny skill.

I carefully worked down the comb for more drop, and the cheek side of the stock for the additional cast-off, using fine sandpaper. The wood was as dull and dark as the old leather trunk case, not from abuse but from overuse of gun oil that had discolored the stock. I can still hear Dr. Norris saying: "Oil never hurt metal."

In my process of sanding I cut into the clear unstained wood of the stock. I finished it with 22 coats of Tru-Oil rubbed to a "London dull" sheen. The old stock seemed to take fire with the sublime smoky black flame of French walnut, formed where the roots join the trunk. I have not seen a London Best stock to surpass it.

When I inherited the Purdey 25 years ago, it was difficult to obtain the short British cartridges, living where I do. I had the chambers lengthened to accommodate 2¾-inch shells. A modern 3 dram equivalent, 1 ounce load is within the bounds of the recommended 40 grains Schultze or 33 grains E.C. loads for which the gun was regulated with 1 ounce of shot. Heavier loads used for proofing these guns represent a margin of safety, not recommended loads. I have shot hundreds of 3 dram, 1⅛-ounce loads in the gun, which is the "maximum service load" of shot indicated in the proof marks.

If, as has been accused, I "butchered" my Purdey, I achieved a gun that performs like an extension of my nerves, vascular system, and brain.

Since I have had it, Purdey No. 21390 has been shot in remote wild valleys, on rugged mountainsides, and on high windswept plateaus not unlike the Scottish moors of other Purdeys. The day after the "Glorious Twenty-Fifth" of October, when traditionally the woodcock flights come through, Kay and I were in a covert where we could expect both 'cock and grouse. I uncased the gun, assembled it and dropped in shells, bringing the stock up to the barrels, thinking as I always do of Dr. Norris at such times. Belton and Quest, unhampered by meditations, were watching me for my signal and I waved them into dense thorn cover. Kay and I pushed after them.

The blank for my stock had been cut from the root of a walnut tree that began to grow in France around 1700. The rough blanks are seasoned for six years before they come to Purdey's where they are kept for five or more years in the factory. To select outstanding patterns in the wood, the blanks are given a coat of stock oil, which brings out the configuration in the walnut.

At the upper end of a woods road festooned with grapevines, I paused to substitute a woodcock load in the right barrel before entering cover along a small run. As I broke the breech, the top lever opened without resistance and hung loosely like a fractured arm.

I closed the gun and tried mounting it, holding the lever in place with my thumb to determine if I could shoot in that manner, but there was no way to be certain my thumb might not slip and let the self-opening breech separate when the gun discharged. Kay located a heavy rubber band in her camera case and I looped it around the trigger guard and up over the top lever with enough tension to hold it closed. The dogs were already working cover and I followed.

Belton pointed on the bank of the little stream. Kay was kneeling beside me for a photo of the point, and Quest swung in and honored Belton. Everything was perfect except my performance. The woodcock didn't give me much, going through a hole in the alders, and I didn't make much of it, but getting off a shot with the rubber band in place was something.

Back home I phoned a gunsmith and without taking time to eat, we were soon driving over the mountain to his shop, still in our shooting clothes.

Purdey No. 21390 and Purdey's The Guns and the Family by Richard Beaumont. All three craftsmen whose initials appear on the shotgun are mentioned in Beaumont's book.

Melvin Forbes had been recommended by my old gunsmith when he retired, and we found him at his bench, surrounded by the clutter and filings that are part of a gunsmith's shop. Kay and I watched him select a turn-screw from a pile and begin to strip the Purdey action, starting with the trigger guard, then the under assembly and both sidelocks, spreading the parts over an oil-and-graphite blackened deerskin on his workbench.

For those of us who grew up with American shotguns, the feel of a London Best is a new experience with the nice balance between the gracefully slender straight "hand" of the stock and the small splinter fore-end.

During the years I had shot my gun I had never seen its viscera. Off the gun, the wafer-thin sidelocks looked smaller, the mirror-polished high carbon steel parts fitting the space with jigsaw-puzzle precision. After seven decades of shooting, the locks were immaculate.

Although Harry Lawrence had said the actioner had put his initials on the inside of the action, I never expected to see them. They were there: LSD on the inner surface of one of the sidelocks, stamped by the British craftsman who had fabricated this piece of jewelry 71 years ago.

My thoughts were interrupted by Forbes's words: "Here's your trouble." He was holding two segments of a small V-type leaf spring, broken near the base of one prong.

That spring was entitled to metal fatigue. It had been flexed each time the gun was assembled, the thousands of times the gun was loaded, each time it was opened to eject a shell, each time the barrels were removed when the gun was cleaned.

Forbes was selecting a strip of high carbon steel and comparing it with the thickness of the broken spring. "We'll make one," he said, moving down the aisle of the shop.

For four hours Kay and I watched him work the strip of steel — tooling, heating, forging a duplicate of the broken spring hand-wrought in the same manner early in the century. Exchange his plaid shirt and blue jeans for a collarless white shirt and bib apron, and you might be seeing one of Purdey's craftsmen leaning over a vise with the same intent expression.

Partway through the evening Forbes went into his house and returned with a plate of breast meat from a wild turkey he had shot east of Allegheny Mountain two days before. Kay and I ate it ravenously.

The piece of metal, well formed by now, was being heated to a cherry red, then immersed in water and tempered. In the final process it was submerged in a trough of burning oil, then dropped in dry lime to cool slowly. At last, at 11:30, after finish work with a file, Forbes laid the new spring in my palm like the gem it was.

While he was installing the action on the stock, I was able to see the exquisite fit of wood to metal, the interior tooling in the walnut with small spurs protruding to fit openings in the action, the dark wood still crisp after all those years. Melvin Forbes carefully assembled the gun, and Purdey No. 21390 was once again whole, only a few hours after the day's breakdown.

The new spring enabled me to be back in coverts without delay, but beneath the relief, I felt a curious need to regain total integrity in the gun with a spring from Purdey's. I placed my problem in the hands of Richard Beaumont, chairman of Purdey's. In my letter I mentioned the initials inside the action and requested a rundown on the workmen who had made the gun. In less than two weeks I received a Purdey top lever spring, together with the following:

The initials A.F. on your fore-end are those of Alfred Fullalove, and you will find reference to him in my book as being the man who trained Harry Lawrence during his apprenticeship. The initials L.S.D. are those of Davidson, and he was the man who told off Tom Purdey for coming to the factory in an intoxicated state! It is very lucky that you have managed to acquire a gun made by three people who I mentioned in the book!!

Thank you so much for writing and I do hope that this spring arrives in good order.

Yours sincerely,
Richard Beaumont

The third man was William Hill, the barrel filer whose initials had been identified for me by Harry Lawrence. I later learned from Richard Beaumont that the man who made my stock was Alfred Dean.

So now I have the missing parts to the story of my Purdey — the men who made the gun. They worked on Purdey guns from pre-World War I until just before World War II. During this period guns numbered 20,000 through 25,000 were built, and I feel fortunate that mine is one of them.

Richard Beaumont's *Purdey's, The Guns and the Family* (David & Charles Ltd., London, 1984) is rich in information about the people and the guns. I went back to it to read about William Hill and "Tommy" Davidson and Alfred Fullalove. Purdey craftsman have always worked under the old guild system, from apprenticeship to a specialized skill, which each man would follow for life. The firm has records of which men worked on each gun.

Harry Lawrence was serving his apprenticeship under Alfred Fullalove at the time Fullalove was working on No. 21390. I feel the touch of hand, for no doubt Harry Lawrence did some of the preliminary work on the ejector mechanism.

To me a gun, like a dog, is dearer for being old. My Purdey is more than a London Best with all the trappings of tradition. It is important to me that it has never been sold, other than by Purdey's; it came to Dr. Norris as a gift from Lynford Biddle; it came to me as a gift from Dr. Norris; it will someday go to another gunner as a gift from me. Purdey No. 21390 has bestowed pleasure and values on three of us so far. I contemplate what it will do for that other man who will shoot it handsomely over gun dogs worthy of the gun.

For many shooting men Purdey's is a realm to be roamed in imagination, a land where there is no compromise in quality, a place that has existed in shining glory. That glory has lasted for 172 years.◆

Dad's
PUMP GUN

There's a whole bunch of sportsmen who declare the Model 31 as the finest pumpgun ever built and will arm-wrestle you to emphasize the point. My father was one.

MICHAEL McINTOSH

Among the things my father gave me is a package made up of chilly November days, a distant sound of dog bells, corn stubble crackling underfoot, and the stiff stride of canvas pants — all packed in a layer of quail and pheasant feathers. It smells of a clear, cold prairie wind and red paper shells, warm from the gun.

From his gun. The one in the cabinet across the room from me now, distinctive in present company for its single barrel and long, flat-sided frame. It's the last one of the group that anyone notices, and it probably looks out of place to any eye but mine.

Dad bought that gun in 1935, paid another man $30 for it. It had been fired a few times but not many. Fifty years later, two years before he died, Dad's gun came to live with me. It's been fired a few times since but not many.

I remember the first time he let me handle it. I was five or six. The sun wasn't up yet, but we were, Dad to hunt quail, I to see him off. The gun was lying on the kitchen table, next to a long canvas case. He picked it up, reached for the case, and must have noticed the way my eyes followed the gun. He slid the action closed, opened it again, glanced inside, and sat down. When he handed it to me, he didn't say "Don't drop it," or "Watch where you point it," or any of the things he might have said under the circumstances. He simply held it out to me and said, "This is a Remington Model 31."

If there ever is a contest to choose the All-American Gun, two of the three finalists will be pumps, and the winner will be the Winchester Model 12. If the whole thing were based on name-recognition and every gun produced were worth a vote, chances are the Model 31 Remington wouldn't even earn an honorable mention. With only about 190,000 built, it never was a plentiful item. But to a certain coterie of shooters, devoted to the pump gun and keenly appreciative of finely tuned machinery, the Model 31 would be the only choice. That, of course, was Remington's intent, though it never worked out as well as it might.

The Model 31 outshines the more popular Model 12 in one important design. A Model 12 comes apart with the barrel, magazine tube, and fore-end together as a single assembly; in the Model 31, only the barrel detaches from the receiver, which makes owning extra barrels a far more economical matter.

By the middle of the 1920s, Remington owned a goodly share of the market for repeating guns. Its Model 11 autoloader, adapted from John Browning's original Auto-5 design, was a particular success, far outselling any similar gun that arch-rival Winchester could put up against it. But Winchester had the Model 12 and with it, a virtual corner on the pump-gun world.

The Remington pumps of those days — the Model 10 and Model 17 — were clever, complicated, and relatively expensive to build. They were no match for the Model 12, neither mechanically nor in the marketplace.

At first, Remington apparently believed that revisions in the Browning designs would turn the trick. Crawford C. Loomis, Remington's principal designer and the man who a few years later would create the great Model 32 over/under, filed patent applications in September 1926 and June 1927 for improvements to "a firearm of the type disclosed in U.S. patent to J.M. Browning #1,143,170," which Remington was manufacturing as the Model 17. The first application asked for, and ultimately received, patent protection on ten different aspects of the "fire control mechanism," which covers virtually the entire trigger, sear, hammer, bolt, and locking system. The second design focuses on the shell carrier and comprises 13 patentable features.

Browning's own patents were so carefully written that Loomis's applications remained under scrutiny for four years before they finally were approved on June 17 and December 23, 1930.

In the meantime, Remington Arms was growing anxious about its standing in the pump-gun market. As a stopgap measure, it had revised the Model 10 somewhat and reissued it in 1929 as the Model 29. By 1930, it must have been clear that revised Brownings weren't going to fare any better than originals and that the new gun would do well to look as little like its predecessors as possible.

When it came on the market in 1931, the Model 31 met those two needs and more. Its mechanical debt to Browning is minimal, though no pump gun can ever entirely avoid Browning's influence. Its appearance, particularly in the long, graceful profile of the receiver, defined a look that Remington guns have had ever since.

From the beginning, the Model 31 was built in 12-, 16-, and 20-gauges, each on a receiver scaled to size. Like the Winchester Model 12, it was designed as a take-down gun, but in this the Model 31 did the Model 12 one better. When a Model 12 comes apart, the barrel, magazine tube, and fore-end remain together as a single assembly; in the Model 31, only the barrel detaches from the receiver, and that makes owning extra barrels a far more economical matter.

The standard Model 31 holds five shells, four in the magazine and one in the chamber. It has a plain barrel, an uncheckered, half-hand grip, and a grooved fore-end. The standard 12-gauge barrel is 30 inches, 28 inches in 16- and 20-gauges; 26-, 28-, 30-, and 32-inch barrels, choked cylinder, modified, or full, were available in any gauge.

As with most of its guns, Remington created a multitude of Model 31 submodels, most according to minor variations in barrels, stocks, and accessories. The standard gun was called Model 31A, and it sold for $39.50. A checkered grip and fore-end raised the price to $42.50. For the police and military markets, there was a 31R riot gun, in 12-gauge only, with 20-inch barrel.

High-grade 31s, like other Remingtons, included 31B Special, 31D Tournament, 31E Expert, and 31F Premier grades. These differed mainly in the quality of wood and the amount of engraving. Some were lavishly decorated: none exist in great numbers.

The plainer guns essentially were standard models set up for specific uses. The 31TC Target Grade, which Remington described as "the finest trap gun ever offered at a moderate price," was built in 12-gauge only and fitted with a 30- or 32-inch vent-rib barrel, a trap-dimension buttstock with full-pistol grip, and a long beavertail fore-end. The rib was forged as an integral part of the barrel — an incredibly expensive operation in a gun that sold for only $108.

The 31S Trap Special offered the same gun with a solid rib, half-hand grip, and standard fore-end. The 31H Hunter's Special differed from that only by its slightly shorter stock with more drop at the comb.

The 31 Skeet, available in three gauges, came with a 26-inch, solid-rib barrel and beavertail fore-end, all for $56.45. An optional vent rib cost $12.40 extra.

Among the more interesting options for the Model 31, Remington offered specially bored 30- or 32-inch 12-gauge barrels meant for long-range shooting. They were available for the Model 11 autoloader and for the Model 29 pump gun as well. Exactly when these first were marketed isn't clear, nor did Remington ever advertise them heavily. They were stamped LONG RANGE and could be had for $3.20 more than the price of a standard gun or barrel. According to my friend Albert Uhl, who has been particularly fond of Long Range Remingtons since the 1930s, these barrels were over-bored to .745 inches, which is about halfway between 12- and 11-gauge, and were given 40 points of choke.

The Model 31's appearance, particularly in the long, graceful profile of the receiver, defined a look that Remington guns have had ever since.

Remington described the Long Range barrels as specially designed for the Remington trap and handicap trap loads, noting that "distance is annihilated." Those who've shot them tend to agree.

In the original Model 31s, a spring-loaded plunger mounted in the magazine tube locks the barrel in place. This was redesigned in 1934 as a threaded plunger, which afforded more security in fastening the barrel to the rest of the gun.

All told, the Model 31 was built in three separate series, each with a few redesigned parts. The first, in production from 1931 to 1936, comprises 12-gauge guns up to serial number 25000 and 16- and 20-gauges numbered 500000 to 519600. Guns in the second series, produced from 1936 to 1940, lack the barrel collar of the original design; these include 12-gauges numbered 25001 to 35000, 16s and 20s numbered 519601 to 530000.

The third series guns, built from 1941 to 1949, are different enough from those of the first two that a good many parts are not interchangeable. The barrel-lock mechanism is smaller, the safety button larger. The hammer, extractor, action-bar lock, trigger lock, and trigger housing are all different.

Sometime during this period, Remington developed a lightweight version of the Model 31, built on an aluminum-alloy receiver. Its black, anodized finish isn't as handsome as the standard gun's rich blue, but it handles beautifully. It's also one of the precious few 12-gauge pump guns that you can carry all day without courting a double hernia.

Remington called the Model 31 *The Gun With the Ball-Bearing Action*, a description not entirely the fruits of some copywriter's fancy.

The action doesn't really ride on ball bearings, but it feels as if it does. Even a gun that's scarcely been fired feels that way, and with only a little use, a Model 31 is as silky as a pump gun can be. Dad's old gun hasn't had a thorough action-cleaning for at least 25 years, and it's far from worn out — but if I cock it, stand it in the cabinet and touch the release button, the action will slide fully open under its own weight, all with a sound as whispery as a fine blade on a hard Arkansas stone.

There is no question that Remington meant the Model 31 to take a chunk out of Winchester's Model 12 sales, and it might have done that if the two had got off to an even start. But by 1931, the Model 12 had nearly 20 years' lead, and there was no catching up. Ironically, it was the Model 31's successor that ultimately gave Remington the upper hand in the pump-gun market.

In the late 1940s, Remington moved quickly to adopt automated manufacturing technology developed during the war and at the same time began designing a new pump gun, one better suited to such techniques. In 1949 the Model 870 came on the market, and the Model 31 went out of production. It was a far-sighted decision on Remington's part. By the time Winchester responded in kind, nearly 15 years later, Remington had all but taken over.

The 870 is no slouch of a gun, and in economic terms it accomplished what the Model 31 could not, but I doubt anyone is willing to argue that it has the same appeal. On the other hand, there's a sizeable population of those who declare the Model 31 the finest pump gun ever built and will arm-wrestle you to emphasize the point. My father was one.

Maybe they're right. I don't have as much objectivity toward that question as I might, especially when I pick up Dad's gun, run a snap cap through the action, and swing on the first pheasant I ever shot, as clear in my memory now as it was 30-odd years ago. As is the look I saw that day in my father's eyes.◆

ITHACA DOUBLES
Overlooked Classics

For almost 70 years, Ithaca made dependable and respected side-by-sides. From field grade to the finely engraved higher grades, they were among the best American doubles.

PAUL RUNDELL

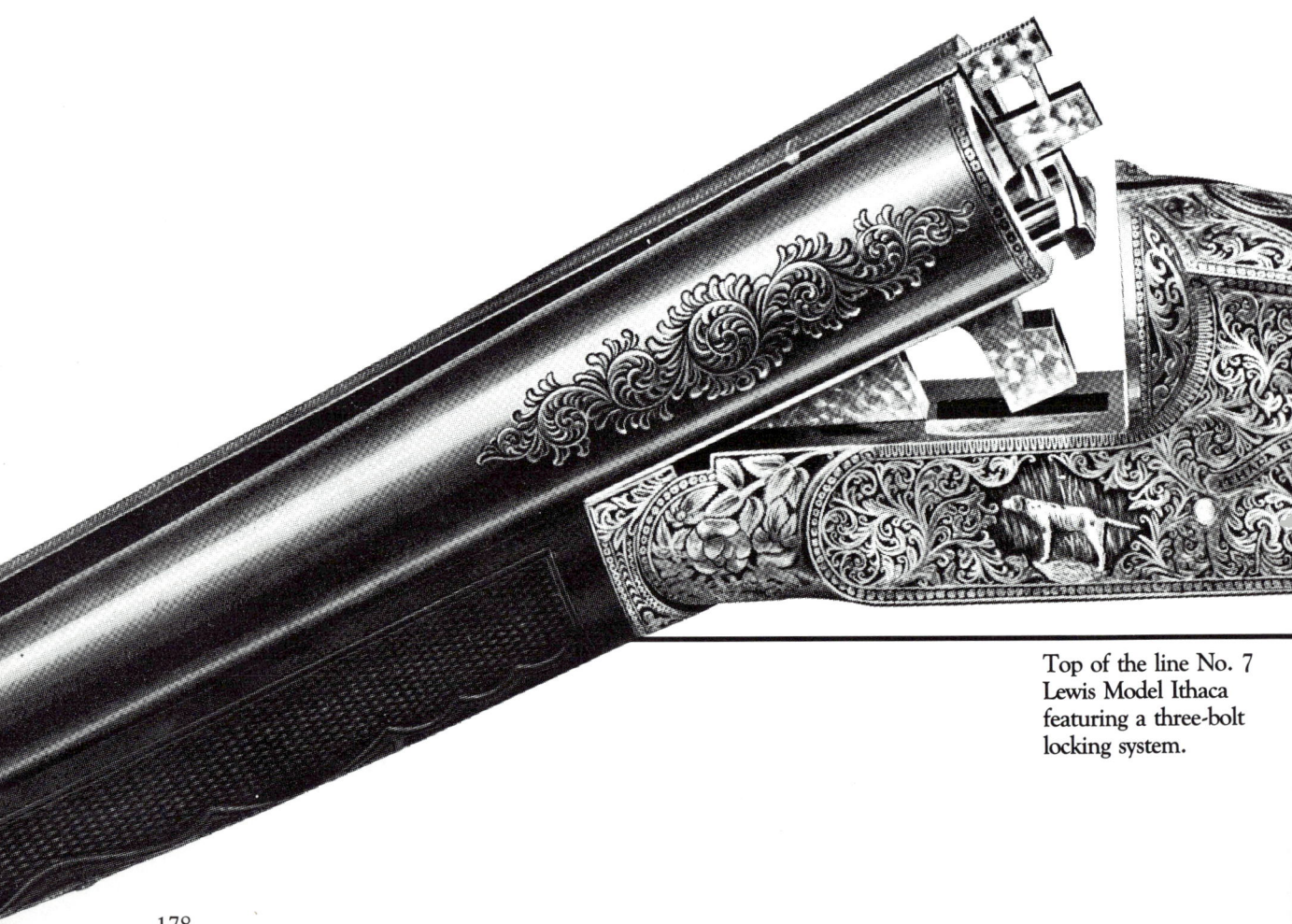

Top of the line No. 7 Lewis Model Ithaca featuring a three-bolt locking system.

Like many enterprises of the time, its origins were humble. It began in a rough little wooden structure perched precariously on the steep gorge of Fall Creek near Ithaca, New York. The virtue of the location was its abundant water power, and the building had already been a mill and wagon spoke shop by the time it was purchased to build guns. In 1880 the first of a distinguished line of Ithaca double shotguns emerged from the dim, oil-lighted recesses of the tiny factory.

During the next 68 years the company grew and the name Ithaca became synonymous with double barrel guns of high quality, sturdy construction and careful workmanship. In an unrelenting quest for improvement and in an effort to provide the American hunter with the best possible gun at reasonable prices, Ithaca doubles underwent five distinct changes in design before shifting market conditions and rising manufacturing costs forced the company to discontinue their entire line of doubles. By 1948 Ithaca had joined other respected American makers of fine side by side guns as casualties of the times.

Leroy H. Smith was the force behind the Ithaca Gun Company. A successful businessman from Lisle, New York, he became intrigued with the possibility of manufacturing a shotgun designed by W. H. Baker, a local gunsmith. A skillful workman, Baker had earned a good reputation for his handmade guns. His earlier designs included a three-barrel gun with two shot and one rifle barrel, but it was Baker's design for an improved double-barrel hammer shotgun that impressed Smith.

Baker's design departed from convention. It was a boxlock action with hammers mounted at the extreme rear of the receiver, in contrast to the usual practice of employing sidelocks on

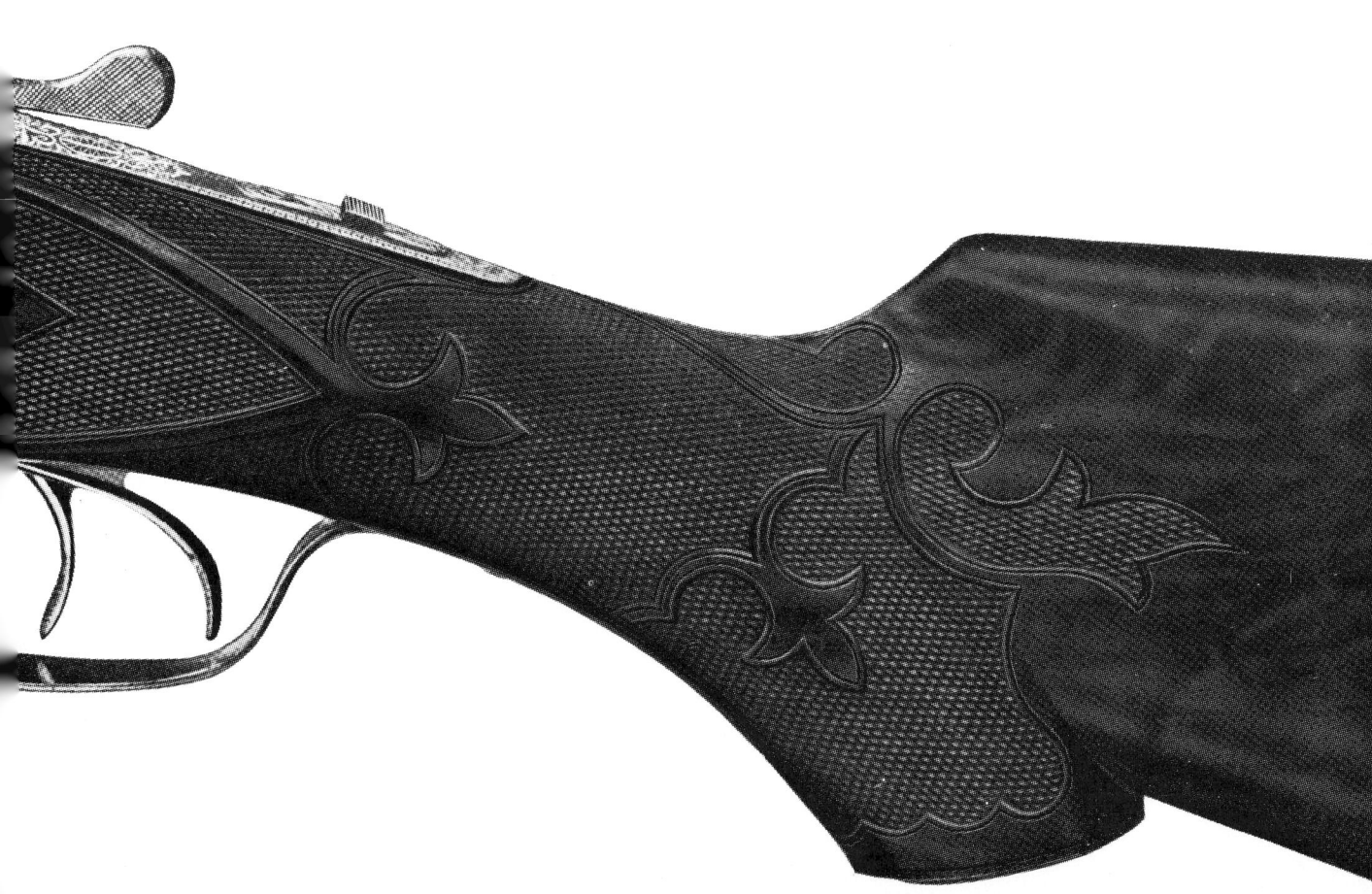

external hammer guns. The Baker double locked with an underbolt and a doll's head extension on the barrel rib. Equipped with the design and $1,800 in capital, Smith set out to produce the Baker double. The Ithaca Gun Co. was born.

The Baker Model contributed substantially to Ithaca's early success and reputation as a maker of sturdy doubles. Early advertisements proclaimed the Baker Model Ithaca to be the "Strongest, simplest and best American gun manufactured."

Nevertheless, among American shooters the trend to hammerless doubles had begun in earnest. In 1892 Ithaca introduced a hammerless gun designed by Fred Crass, a toolmaker employed by Ithaca. Production of the Crass Model Ithaca began at serial number 17,235 and continued through number 94,108 in 1903, the final year of production.

Damascus or twist barrels were standard for the times, but along with the trend to hammerless guns, a shift to the use of stronger fluid steel for barrels occured during the years just after the turn of the century. Company sources place the introduction of fluid steel barrels on Ithaca guns around 1902, and the 1903 Ithaca catalogue offers fluid steel barrels in addition to stub twist and Damascus.

The transition to fluid steel did not take place overnight, for many shooters still preferred the old construction, and Ithaca continued to offer twist or Damascus barrels on their guns for a number of years after the introduction of fluid steel. The 1916 catalogue contains the last mention of the availability of Damascus and stub twist barrels.

If one were to analyze models by barrel type during these transition years, you would find that some late manufactured Crass doubles were produced with fluid steel barrels and that the subsequent models of the Ithaca line — Lewis, Manier, and Flues — were barrelled both with modern fluid steel and with Damascus or twist. As time went by and demand for Damascus declined, it seems likely that an increasing percentage of these Ithaca models were fitted with fluid steel tubes.

Earlier Ithacas having barrels of modern steel are sometimes found. As late as the 1920s and 30s, the company would fit new steel barrels to their guns to replace the original twist or Damascus if the customer desired it.

In 1904 the Lewis Model Ithaca was introduced at serial number 94,109. Although the cocking mechanism remained essentially the same as the earlier Crass Model, the Lewis was the first Ithaca advertised as locking with a three bolt system. Ithaca catalogues of the Lewis era show the locking arrangement employing two separate cuts in the barrel extension plus an underbolt.

By this time the Ithaca line had become extensive. Hammer guns continued to be offered for those who preferred them, and there was a model of hammerless gun to meet virtually every need, taste, and pocketbook. The 1904 Ithaca catalogue lists no fewer than nine hammerless models including No. 1 Special, No. 1, 1½, 2, 3, 4, 5, 6, and 7. All guns were available in 10, 12, and 16 gauge. Prices ranged from $37.75 for the No. 1 Special with Nitro Steel Barrels through $100 for a No. 4 with "fine imported four-blade Damascus, Crown or Krupp fluid steel barrels, fine selected French walnut stock, rolled gold triggers, all metal parts beautifully engraved by hand with dogs, bird and game scenes."

At the top of the line was the No. 7 Ithaca, described from the catalogue as: "The best imported Damascus or Whitworth Fluid Steel barrels money will buy, the best Italian walnut

---

The most elaborately finished and expensive guns ever produced by Ithaca were the famous Sousa-grade guns, in honor of the great band leader and composer, John Philip Sousa. They were made in both single barrel trap and double barrel models.

stock and fore end to be had, full pistol grip unless ordered, engraved skeleton butt plate, the most elaborate checkering and engraving, gold triggers and name plate, a gold dog inlaid on each side of frame, inlaid gold bird on bottom of frame; in fact, the best of everything and unequalled by any American or European guns at any price. 10, 12 and 16 gauge . . . List Price, $300.00."

The Manier Model Ithaca, incorporating a different cocking mechanism than its predecessors, was introduced in 1906 at serial number 130,000. Somewhat shorter lived than the Lewis Model, the Manier was manufactured for only three years. In 1908 it was replaced by the Flues Model.

Emil Flues was a gunsmith employed by Ithaca to do research and development work. The Flues Model of 1908 introduced a refined bolting system incorporating tapered locking bolts, although the three-bolt design introduced with the Lewis Model was retained. The 1909 Ithaca catalogue proclaimed the Flues to be "Our Improved Model," and asserted it to be a gun that "will stand a lot of use and misuse." The Flues proved to be a durable and long-lived model. Production began in 1908 at serial number 175,000 and ended in 1926 at serial number 398,365.

The Flues years witnessed the beginnings of the most elaborately finished and expensive line of doubles ever produced by Ithaca. These were the famous "Sousa-grade" guns, a designation bestowed by the company in honor of the great band musician and composer John Philip Sousa. A dedicated trapshooter, Sousa regularly participated in national competitions. In 1918 he ordered a Flues Model single barrel trap gun with a specific request for a great deal of engraving. Ithaca's master engraver Bill McGraw was re-

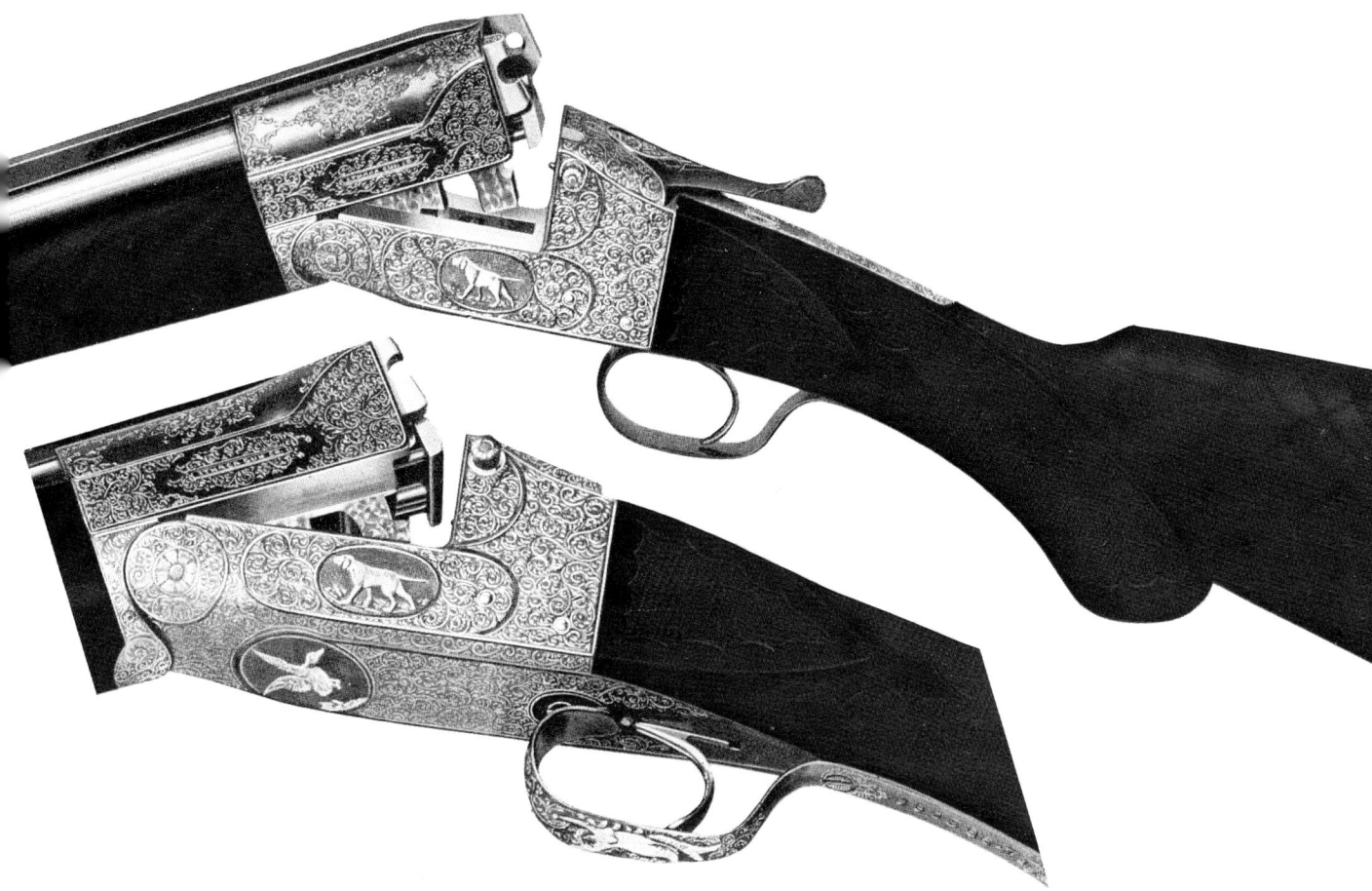

sponsible for embellishing the remarkable shotgun. The action was elaborately scroll engraved and at Sousa's request a buxom mermaid inlaid in gold graced the trigger guard.

Sousa-grade Ithacas were made in both single-barrel trap and double-barrel models. Typically, Sousas had a gold pointer and setter inlaid in bas-relief on either side of the action and sometimes an eagle or flying duck underneath. The gold mermaid, hallmark of the Sousa Ithacas, was inlaid on the trigger guard.

In the years which followed, a small number of Sousa-grade doubles were produced. Company sources indicate that three Flues Model Sousa doubles were made and 11 of the NID Model (New Ithaca Double), which was introduced in 1925. Twelve-gauge Sousas were the most popular. Of the NID Sousa doubles, ten were 12 gauge and one was .410 bore.

The 1918 catalogue lists the Sousa double at $500. Prices rose steadily, and by 1934 the "Sousa Special" listed for $812.50.

Some years after the great band leader's death, the Sousa designation was dropped. The highest grade double in the Ithaca line now became known by the price it commanded. Hence, the 1941 catalogue lists the "$1000 Grade Ejector," and by 1946 it had become the "$1500 Grade." As manufacturing costs rose sharply in postwar years, the model went to $1700 and finally to $2,000.

The NID Model represented the final design

The colorful artwork of Lynn Bogue Hunt is featured on this 1919 Ithaca catalogue.

stage of Ithaca double guns. Introduced in 1925 at serial number 425,000, the NID differed significantly from earlier Ithacas in both cocking and bolting systems. The company's first gun designed expressly for modern ammunition, the NID design replaced the "cocking hooks" of earlier models with rods connecting cams positioned in the forward part of the frame to the hammers.

The locking arrangement was also completely different. The underbolt was eliminated, leaving the frame stronger in this area. The barrel extension was redesigned to engage Ithaca's new rotary crossbolt.

The NID Model served as the basis for Ithaca's famous magnum doubles, introduced first in 10 gauge in 1932. Magnum doubles featured a somewhat larger and heavier frame as well as an additional locking lug through the bottom of the frame. Ithaca sources state only about 800, 10-gauge magnums were made, with the remainder of the 1000 serial numbers being applied to magnum 12-gauge guns.

References to the speed of Ithaca locks are found frequently in Ithaca promotional literature of the Flues and NID periods. Ithaca lockwork was rugged and simple. The hammers traveled less than one-half inch and were very fast. Tests conducted at Cornell University confirmed hammer fall time to be only 1/625 of a second. Ithaca advertising often stated that a bird flying at 60 miles per hour would travel only about one inch during that time.

Ithaca doubles were frequently credited with excellent shooting qualities. Company advertising cited skillful barrel boring as the reason and gave much of the credit to "Uncle Bob" Edwards, a gentleman with a flowing white beard who had joined Ithaca after working with Baker. Edwards specialized in boring shotgun

> *Tests conducted at Cornell University confirmed hammer fall time to be only 1/625 of a second. Ithaca advertising stated that a bird flying at 60 miles per hour would travel only about one inch during that time.*

barrels. Referring to the shooting qualities of Ithaca guns, noted firearms authority Elmer Keith wrote in his book *Shotguns By Keith:* "Both the old and the new models were superb shooting guns; in fact we have never seen better patterning guns, nor any guns shooting higher percentages in full choke than the Ithacas."

Over the years Ithaca Gun Co. absorbed a number of gunmaking firms, including the Union Firearms Co., Syracuse Arms Co., Wilkes-Barre Gun Co. and the Parry Firearms Co. The most notable of these was the Lefever Arms Co., acquired by Ithaca in 1915.

The Lefever name had long been associated with a distinguished line of high-quality doubles, and Ithaca apparently assembled some of these guns from parts on hand during the years 1916-17. According to current Ithaca sources, these Ithaca-assembled Lefevers numbered about 1,000.

In 1921 Ithaca introduced a newly-designed Lefever double of simplified boxlock construction. Known as the Nitro Special, it was made in gauges 12, 16, 20 and .410 and sold for substantially less money than the Field Grade Ithaca. Production of the Nitro Special began at serial number 100,000 and continued through serial number 361,199 in 1947. In 1935 Ithaca brought out the Lefever A Grade, a somewhat higher grade gun costing a bit more money. Options such as single triggers, beavertail forearm, automatic ejectors, recoil pads, and ivory sights were available on both models.

Other Lefever doubles were also offered by

Louis Agassiz Fuertes, recognized as one of the best-ever painters of birds, created Ithaca's 1934 catalogue cover.

Ithaca, although apparently only for relatively short periods of time. The 1935 Lefever price list offers a "Model 4 — Lefever Double Ventilated Rib Trap" gun and in 1936 the company listed a "Model 6 — Special Skeet Gun" which was available in .410, 20, 16, and 12 gauges.

From 1929 through 1946 Ithaca also produced the modestly priced Western Arms Long Range Double. Similar in lockwork but differing in frame from the Nitro Special, the Western Arms Double could be ordered with optional auto ejectors and single trigger. In 1930 the gun sold for $20. Auto ejectors cost an additional $8.25.

Ithaca produced many beautifully embellished doubles, and no single man contributed more to this artistry than Bill McGraw, Ithaca's master engraver for 62 years. His inlay work was notable for its use of pink, green and yellow gold, and much of his engraving included scenes of ducks and pheasants.

Probably the most famous of McGraw's creations was the 12-gauge double Ithaca built in 1933 for King Feisal of Iraq. A gift from a group of Feisal's American friends, this gun absorbed the energies of five gunsmiths for more than two months and cost over $5,000. Decorated with embossed and inlaid scrollwork of 24 karat gold, the action bore the jewel-encrusted design of the royal coat of arms. The stock was of highly figured burl walnut and decorated with carving.

Field grade Ithacas are much more numerous than high grade models, and a man wanting a good example of an Ithaca double is more apt to find one of these. Prices have risen dramatically from the old days, of course, when a man could buy a Field Grade Flues for $32.50 (1918) or a Field Grade NID for $49.16 (1941).

Herschel Chadick (*Sporting Classics* firearms columnist), in response to a recent query from the author, put the price of an original NID Ithaca Field Grade with 50-60 percent blue and case colors at $400-$500. Automatic ejectors might command an additional $200. Prices quoted were for 12-gauge guns. Smaller gauges command more money and value accelerates rapidly as condition approaches mint. Ithaca prices have not reached the levels commanded by some other American doubles, but Ithacas are excellent guns, well worth the investment for collecting and shooting pleasure.

Detailed analysis confirms the abundance of Field Grade NID Ithacas in comparison to higher grade models. Company records indicate that only 560 No. 4 Grade Ithacas were produced, and only 22, No. 7 Grade guns. Small gauge NID doubles are also scarce even in Field Grade, for only 295 28-gauge guns were made in this lowest priced model. Ithaca sources show 790 Field Grade NID .410 doubles. High-grade small gauge NID guns are just plain rare.

There is good news for those who own an Ithaca double which may need service. The Ithaca Gun Co. is again working on its old doubles and offers a rather extensive restoration service. Ithaca is equipped to overhaul actions, including the manufacture of necessary small parts, restocking, resoldering ribs, removing dents, refinishing of metal and wood and new case colors. However, Ithaca cannot supply major parts which include barrels, frames, forearm irons and trigger guards. Requests for estimates should be directed to Al Burns, Product Service Manager.

Well-made and dependable, Ithaca doubles met the needs of the American sportsman for nearly 70 years. It is to Ithaca's credit that their reputation has lived on long after the guns themselves have vanished from dealers' racks.♦

CANVAS of Steel

An artist in the medium of steel and gold, Vermont gun engraver Winston Churchill creates elegant masterpieces on the finest of firearms.

DAVID E. PETZAL

If you want to call it by its proper name you can call it 4140 chrome-moly. It is the steel most often used in the making of guns and, when properly heat treated, it can withstand pressures of up to 150,000 pounds per square inch. It is one of the toughest, most unyielding substances in the world...and it is also one of the greatest artistic mediums we have.

In this year of 1984, there are perhaps 100 places you can go in the United States to find a person who will sit down with a few steel gravers, a hammer or two and a ball-shaped vise, and engrave a design of your choice on the firearm of your choice. Some of these engravers are butchers. Most are skillful. A few — a very few — are incomparable artists. There are perhaps two or three names that fit into this category, and the one that belongs indisputably is that of a 44-year-old Vermont Yankee, Winston Gordon Churchill.

Jim Carmichel, Shooting Editor, *Outdoor Life* magazine:

"If you were to compile a list of the top engravers in the world, you'd come up with Winston and four Italians. He is a unique talent, in that he's mastered so many different styles of engraving.... He can be the greatest American engraver of all time."

Bill Ward, president of Griffin & Howe, New York, New York:

"Winston's great strength is his ability to reproduce a scene in metal almost perfectly. In this technical sense, there is no one better. Also, time means nothing to him. If it takes 100 hours to engrave a scene the way he thinks it should be done, he'll take 100 hours, not 99...."

David Miller, president of the David Miller Company, Tucson, Arizona:

"Winston is one of the two best engravers in the United States right now, which is to say one

Gold inlaid sideplate of a Fabbri shotgun recently engraved by Churchill. He considers it his best work.

of the two best in the world. The other is Lynton McKenzie. I don't think anyone can equal Lynton's scroll work, but no one can match Winston's scenes. He does gold inlays that are just unbelievable."

"Winston's subjects are always appropriate, as contrasted to some of the great Italian engravers who are very skillful, but who will stick anything on a gun, even if it doesn't belong. He takes tremendous care with his backgrounds. He showed me one scene where there was an oak tree in the background, and underneath it, he had engraved leaves of different types, because that's the way it is in nature. Leaves blow in and they mix."

Gary Herman, founder and director of Safari Outfitters, Ridgefield, Connecticut:

"I think that if you take into account Winston's execution, he's unquestionably the top engraver in the United States. No one can touch him. If you want to rank him in the world, that's more difficult, because different countries have different styles, and you really can't compare them, any more than you can say that one particular car is the best in the world. But if you had to narrow it down to the best three, he'd certainly be among them."

Churchill was born on December 19, 1939, in Cavendish, Vermont, and went to high school in Springfield, where he spent as much time in the shop as in the classroom and became an apprentice patternmaker in 1958, the year that he graduated. In his patternmaking classes, he met a man who was to have a profound influence on him — Dave Farr, who, in addition to his woodworking expertise, was also a gunsmith.

After graduation, the Navy claimed Churchill for four years and put him to work carving mahogany presentation plaques. Discharged in 1962, he returned to Vermont and held a variety of jobs, including a stint as a photographer for a mapmaking service, and a term in the machine shop of a General Electric plant in Ludlow, Vermont. All this time he was learning the

These sideplates on a Browning gun reflect Churchill's amazing ability to create game and hunting scenes.

gunsmith's trade from Dave Farr, and built his first (and only) rifle, a Mannlicher-stocked .30/06 Springfield.

In 1967 Farr introduced him to Robert Kane, an engraver who lived in Newfane, Vermont. Kane saw promise in Churchill, and gave him some simple engraving work to do at home. As Churchill put it:

"That was really the start. I liked engraving, and I got hold of L. D. Nimschke's pattern book and Prudhomme's *Gun Engraving Review* and started working by myself. I worked on a .22 rifle, which was my first engraved gun. The Prudhomme book was great because it had all sorts of work, good and bad, and you could pick out what was good and what wasn't.

"I didn't last with Kane. He wouldn't let me sign what I did, and he finally hired a high school kid who he could pay less. I had to keep working because I had a family. I did anything I could, even digging ditches.

"In 1967, on my vacation, I had gone down to New York City to meet Joe Fugger, the great engraver who worked at Griffin & Howe when it was owned by Abercrombie & Fitch. I was one scared country kid when I went in there. A salesman told me that Joe wouldn't see visitors under any conditions, but I must have looked so miserable that he let me in to see Leo Martin, who was the manager of the A&F gun department.

"Leo was on the phone, chewing a piece out of someone. I'd never heard anything like that in my life. I thought, 'My God, when he gets off the phone he's going to chew me up,' but when he put down the receiver, he gave me a big smile and said, 'What can I do for you, son?'

"So I got to see old Joe, and they told me to keep working and come back sometime. Finally, in 1969, I did. I never dreamed I'd work in New York City, but they talked me into moving to New Rochelle and working at Griffin & Howe, doing checkering and engraving. I couldn't have been luckier. Griffin & Howe was a

wonderful place to be. There was a ton of checkering to do before I could engrave, but I got to work with Old Joe and to see a lot of the great engraving from different artists that came through G&H.

"I can't say Joe really taught me anything as far as technique is concerned. I just watched. You can learn more from watching for five minutes than you can by reading two books. But every day we'd eat our sandwiches together, and he'd talk to me about engraving, making sketches on pieces of sandpaper and napkins. I don't think he ever drew on a regular piece of paper.

"What I did get from Joe was the philosophy I've used ever since — quality. Do it right. Don't skimp. Don't be sloppy. That, and the tremendous emphasis he put on backgrounds and the detail in them. The other thing that Old Joe did was to put me on my own creatively. I remember the first time I had to engrave a pistol grip cap. I went to him and asked what he thought Griffin & Howe would want, and he said, 'Don't ask them what they want! You tell them what you're going to do.' He was always like that; he forced me to do my own designing."

Churchill worked day and night during this period. Every morning he would carry his 20-pound engraver's vise (a vise mounted in a rotating ball that can be set at any angle, and which looks like an extra-large, chrome-plated shotput) to work, and carry it home at night in a leather bag he made for it. He got better and better, always learning.

... *What drives me crazy is the man who thinks of the gun in terms of 'I paid X dollars for it in 1984 and I'll get X dollars for it in ten years.'*

In the fall of 1973, he left Griffin & Howe and returned to Ludlow, Vermont. By then his reputation was growing, and he had a small clientele. In addition, he did free-lance work for Griffin & Howe. In the years that followed, his fame grew. He engraved knives and handguns and stocked and engraved rifles and shotguns.

It was in the late 1970s that he began a practice that has gotten him almost as much notoriety as his work. He charged what was then very large sums of money for his talents.

Prior to that, engravers, like other artisans who worked with guns, were scandalously underpaid. In years past, engravers had, for the most part, been factory workers and were paid on the same basis as the men who put bolts together or packed the guns for shipping. Many of them were European trained, products of the apprentice system which stressed years of learning and a lifetime spent at low wages. A great artist like Josef Fugger would complete a major work and see the most of what the customer paid for it go to Abercrombie & Fitch. It was the way things were done.

So the stories began circulating about the really substantial amounts of money that clients were spending on the work of the engraver with the Prime Minister's name. They were fair prices; they were what was deserved, and today they are not the rule, but they are far more common among top craftsmen than they were ten years ago. Gary Herman sums it up very well:

"Some people say, 'Yes, Churchill's great, but his prices are out of line.'

"That's crazy. The man is dealing with the customers of his choice, getting his prices, doing the work he wants. He's got a two-year backlog with people begging to get on the waiting list. There's nothing crazy about that."

Gradually a pattern evolved whereby Churchill

would take on only a few major jobs per year and a few incidental ones. When he started, he would stock as well as engrave a gun, but the stocking has gone by the wayside, with the last one completed in 1978. It just doesn't interest him enough. The irony is that his woodwork is a superb as his engraving. Although he stocked only ten rifles and shotguns in his career, he was selected as one of the thirteen gunsmiths featured in Ron Toew's definitive work, *Contemporary American Stockmakers.* Toew, who made an exhaustive study of the subject, ranked him alongside such greats as Al Biesen, Jerry Fisher, Dale Goens and Monte Kennedy.

Churchill will engrave any gun that meets a fairly rigid set of criteria. First, the unengraved arm must have inherent quality; it must be a fine piece of machinery. Examples of this are older Smith & Wesson revolvers, pre-1964 Model 70 Winchesters, unengraved Browning Superposed shotguns, and Ruger No. 1 single-shot rifles (after they have been extensively re-worked to bring them up to his standards).

Second, the gun must be aesthetically pleasing. The toughest job Churchill ever had was a Remington Model 32 which he restocked and engraved soon after leaving Griffin & Howe. His wife called it "the albatross."

"I put everything I had into that gun, and it was still ugly. That hood over the barrels, the trigger guard, the trigger, just ugly. I wouldn't do it today."

You must also bring money. Churchill's present rate is $35 per hour, which is far less than lawyers, psychiatrists and cocaine dealers charge. But it can add up. For example, Churchill recently completed a full-house engraving and inlaying job on a Fabbri shotgun and spent 1,496 hours on the project. If you do not have a pencil handy, this is $52,360 worth

Floorplate and trigger guard of a Mauser rifle.

of engraving and inlay work.

Nevertheless, he is anything but short of customers. There are, Churchill says, about ten people who regularly keep him busy on the big jobs he prefers. He does not like to be committed for more than two years at a time, and substantial fees are a good way to keep things within that limit.

What are his customers like?

"They're all between 35 and 55 and they're all alive. That's the best description I can give you. They're interested in what's happening now. Most important, they're interested in what I do and they appreciate the finished gun as a work of art, not just as an investment. I don't really mind a customer butting in and trying to tell me how to go about a job because I'll ignore him for the most part anyway, but what drives me crazy is the man who couldn't care less about the gun and just thinks of it in terms of, 'I paid X dollars for it in 1984 and I'll be able to get X dollars for it in ten years.'

"Don't get me wrong; I don't mind my guns increasing in value. In fact, I wish that more of them would change hands. But the guys who just want a Churchill-engraved gun for the bucks

Gold inlaid action of a Browning shotgun.

involved ... it's like the people who pay $100,000 for an A-1 Special Parker. It's dead, it's past, there's no life to it. The great work isn't being done in the factories anymore, it's being done by individuals."

When Churchill begins a job, he does indeed listen to the customer and hashes out the major features of the inlaying and engraving. From there, it's largely a matter of improvisation. He will go to magazines and books to look for likely models (he currently subscribes to twenty-six magazines for this purpose).

"Take the covey of quail on the Fabbri. I went through lots of sources to find the individual birds that would comprise that group. I'd get one here, one there, and sometimes I couldn't get one just right in an illustration so I'd draw it myself.

"Once I get the scene designed, I paint the gunmetal with Chinese white lacquer and transfer the basic lines into it. Then I cut the main lines with a hand graver. If I'm not dead sure how I want it to look, I'll do the main lines as a series of dots. You can follow them well enough to envision what a design will look like, and if you want to change, you can engrave right over them, which you can't do with lines. I never have a complete layout in mind when I begin; you just go along and see how it comes out."

The tools are simple. He now does 95 percent of his work with a hand graver. Add to that a couple of vises, a hammer-driven graver or two, and you have it. The eyes and the hands are the basic ingredients. And, of course, the imagination.

"I try to put in a good seven- or eight-hour day on the job. Add to that a lot of time doing photography and looking for ideas. But I can't take all day at the bench. I'll quit, and cross-country ski, or run, or go bicycling. I did a century last summer; that's where you cover 100 miles on a bike in 12 hours. I did it in ten and a half."

Like any top craftsman, Churchill has colleagues whom he particularly admires.

"Among the gunsmiths, Jerry Fisher and David Miller. And there's a fellow named Dan Cullity who lives on Cape Cod who's about the best in the world at restoration. He's got all the old formulas, and there's nothing he can't do with an old gun. Awful nice fellow.

"You'd have to list Creighton Audette, who's a genius at any kind of metalwork, and a grand fellow, and my old friend Joe Sovenyhazi at Griffin & Howe who's a wonderful gunsmith and can tackle any kind of shotgun problem.

"As far as engravers, I like Firmo Fracassi for his detail work, and among the old-time engravers, Gustave Young. He had that same superb quality that old Joe Fugger did and that same wonderful detail."

What does he hope to do in the future?

"Oh gosh, I really don't know. I can't sit down and say to myself, 'Here's what you have to do to be a better engraver.' It doesn't work like that. You pick up a little bit at a time; you can do things a little better this year than you could last year. You really can't answer a question like that."

In 1977, when his house was being built, the studs had just been erected when he sighted along the foundation and yelled for the builder. One stud, he claimed, was out of line. The builder denied it, but a chalkline was produced, and one stud, among all those lined up in the 50-foot foundation, was found to be ¼-inch out of line. It was removed and re-set at Churchill's request.

For Winston is, when all is said and done, a genius second and a Yankee first.♦

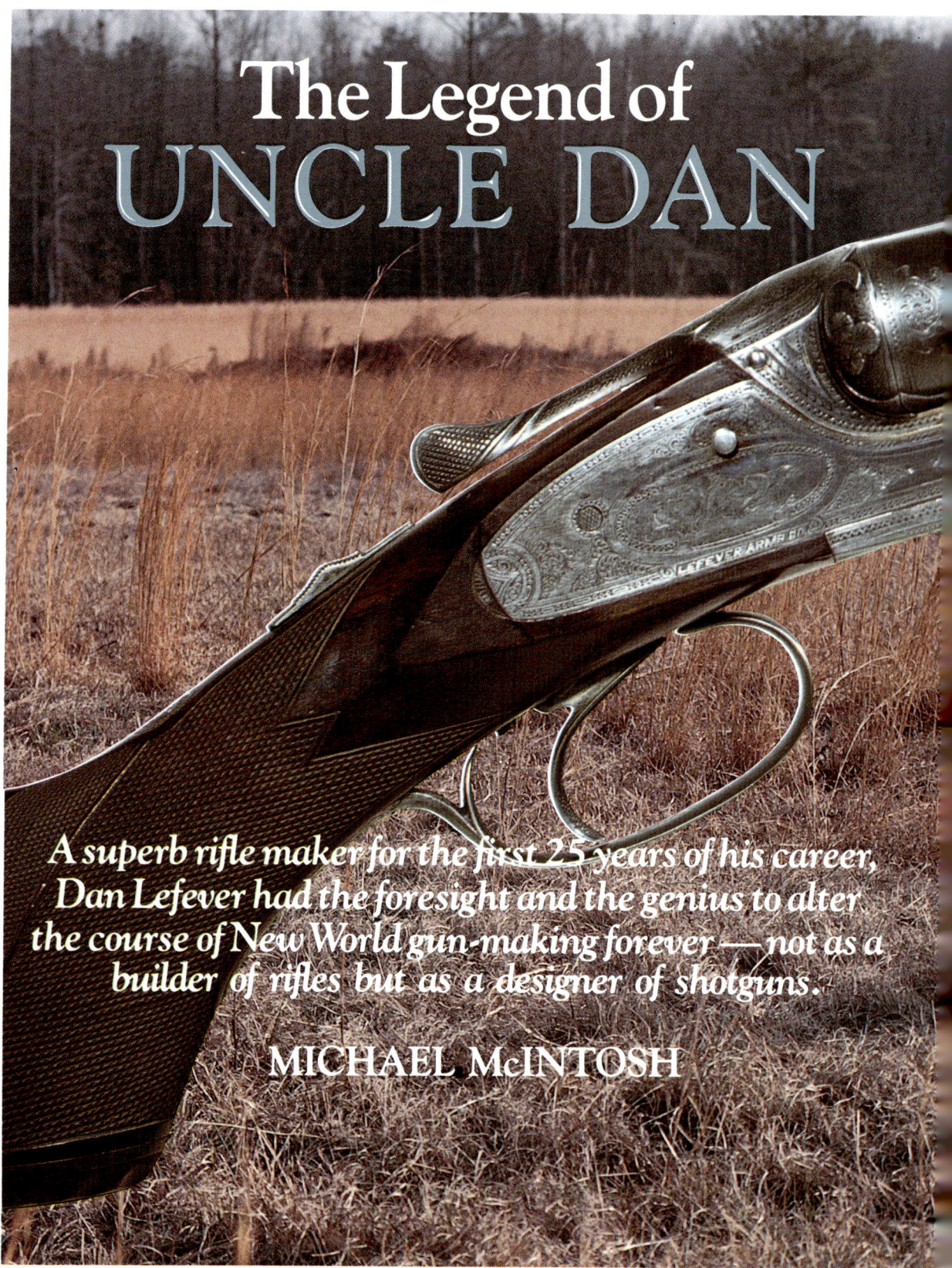

The Legend of
UNCLE DAN

A superb rifle maker for the first 25 years of his career, Dan Lefever had the foresight and the genius to alter the course of New World gun-making forever — not as a builder of rifles but as a designer of shotguns.

MICHAEL McINTOSH

Photograph by Art Carter

Dan Lefever came late to shotgunmaking. In April 1848, when he was 16 years old, his mother bound him out for a five-year apprenticeship with one of the great riflesmiths of the period, William Billinghurst of Canandaigua, New York. The choice of trade must have set well with Lefever; he wrote almost 60 years later: "I walked eleven miles through the snow . . . to learn to make guns."

The Lefever family had immigrated to America in 1690. Of Huguenot origin, they had been driven out of France by the religious persecution of the 17th century. One of the sons, Philip Lefever, established a trade in gun and tool making in 1710 and pursued it successfully for more than 50 years. His son George was an officer of the Continental Army in the American War of Independence. By the early 19th century, the family had settled in central New York, where Dan Lefever was born in 1831.

After his indenture with Billinghurst, Lefever set up his own shop in Auburn, New York, 50 miles east of Canandaigua; for the next 25 years he built rifles almost exclusively.

Almost. In some corner of his busy mind, the image of a shotgun was taking shape, one that would ultimately alter the course of New World gun making. It was a gun that would make Dan Lefever a legend.

Lefever's genius emerged quickly. His little shop soon became famous for turning out beautiful, highly accurate rifles. Lefever was one of the first gunsmiths in America to anticipate the transition of the muzzle-loading gun to a breech-loading design. His excellent work in converting muzzle loaders to breech loaders made it a reality, affording him not only a growing reputation but a spot in the forefront of a new age.

Lefever began experimenting with shotguns

almost immediately. His earliest shotguns were two and three barrel combination guns, built between 1854 and 1870; by 1867 he was experimenting with break-action, breech-loading guns. The demand for rifles during the War Between the States, however, left little time for work with sporting arms. In 1862, the New York State Company of Sharpshooters left Rochester, headed for Virginia. Each man was armed with a fine 'scope-sighted Lefever gun, which each had bought and paid for personally — a fine tribute to Lefever's skill as a riflesmith.

In 1870 Lefever formed a partnership with Francis Dangerfield. Although it was to last only three years, it was a productive one (it seems that Dan Lefever's business alliances were always brief; Lefever was a superb gun designer, not a businessman). Dangerfield and Lefever developed a design for the first top-lever bolting system for the break-action gun; it was patented on September 3, 1872.

A year later Lefever was in Syracuse, New York involved in a gun-making business called L. Barber & Company; nothing is known of the venture, except that it was short-lived. By 1876, he was working with John A. Nichols. From that partnership came a series of superb underhammer target rifles and, finally, a shotgun.

The long day of the muzzle-loading gun was coming to a close in the 1870s. At the 1876 Centennial Exposition in Philadelphia, W. W. Greener won a prize for a hammerless, breech-loading gun built on Theophilus Murcott's 1871 patent. William Anson and John Deeley, representing Westley Richards of London, introduced the world's first hammerless double that was cocked by barrel leverage. Dan Lefever wasn't far behind.

The central problem in designing a break-action gun was finding a reliable means of fastening the barrels to the frame. In earlier experiments, Lefever tried using a rib extension that fit into a slot in the top of the standing breech. By 1877 he had refined his idea and was granted a patent on June 25, 1878. The top-fastener was so effective that it ultimately appeared on virtually every fine double gun built in America — including Lefever, Ithaca, L. C. Smith, A. H. Fox, and the famous Parker doll's-head.

This improved rib extension was the final step in the development of Lefever's first shotgun. It appeared in 1878, and it was the first successful hammerless, breech-loading shotgun made in America.

The gun was exhibited in St. Louis that year, competing against 21 of the world's finest guns. Two gold medals were at stake, one for the best American gun and another for the best gun in the world. Dan Lefever won them both.

This early Lefever was basically a conventional-looking sidelock double. The internal hammers were cocked by a lever on the

The Lefever Arms Thousand Dollar Grade was built to suit any special requirements of the shooter.

left side of the frame, a feature developed by Greener a few years before. The twin locking bolts, one fitting to a notch in the barrel lump and another engaging a slot in the rib extension, were operated by a rocker-type thumb piece on the top tang. It was a graceful gun and wonderfully strong; but for all its excellence, Lefever saw it as only a beginning.

Around 1880 Lefever and Nichols dissolved their partnership. Remaining in Syracuse, he founded Lefever Arms Company in 1884. The side-lever gun was still in production, but a new design was in the making. It appeared in 1885 and was called the Automatic Hammerless. It was a masterpiece.

The inventive design did away with the side lever. The hammers were cocked by barrel leverage, hence the "automatic" part of the name; the thumb-piece latch remained, but the locking was simplified to eliminate the underbolt and relied entirely on the top-fastener. The thumb piece was discarded within a few years in favor of a top lever adapted from Dangerfield's patent.

The Automatic Hammerless is one of the most innovative, most completely adjustable shotguns ever built. Lefever called it the "compensated action," and designed it so thoughtfully that virtually every mechanical adjustment and every type of wear in the action can be corrected with only a screwdriver.

The barrels pivot on a ball-and-socket joint instead of the conventional hinge pin. A heavy, tempered screw, rounded at one end, fits into the front of the frame. The rounded end protrudes into the water-table slot and matches a milled recess in the barrel lump. Should either ball or socket wear down to the point that the action becomes loose, it can be easily tightened by seating the screw slightly deeper.

To compensate for any lateral play that might develop through wear on the sides of the barrel lump, the lump itself is split and fitted with a tapered setscrew. Turning this screw to the right spreads the lump enough to achieve a perfect fit. The fore-end iron has an adjustment feature, also, to compensate for wear in the hinge joint.

This cocking system is typical of Dan Lefever's

genius for design. At the heart of it is a cocking hook that fits into the water-table slot and engages a pin in the barrel lump. When the action is opened and the breech-end of the barrels swing up, the cocking hook lifts the hammers into their sear notches. The leverage, directly assisted by the weight of the barrels, has an extremely high mechanical advantage.

The cocking hook has two additional functions. As the gun is opened, the hook's rounded top cams the extractors to lift the fired shells a quarter-inch out of the chambers. The automatic ejectors are not tripped until both hammers are at full cock, an arrangement which promotes optimum cocking leverage. The cocking hook also serves as a check to dampen stress on the hinge joint should the action be opened with unusual force.

Cocking levers can be synchronized so both hammers reach full cock at precisely the same instant; trigger-pulls are adjustable to as light as two pounds; the safety slide converts from automatic to manual, all by the turn of a screw.

The locking bolt, too, has a compensating screw. Should any wear develop in the top-fastener, a slight turn of this screw brings the rib extension back into a perfect, tight fit against the frame.

Though the Automatic Hammerless is fitted with sideplates, it is not a true sidelock. The locks are attached to the frame rather than to the plates. When detached, the sideplates provide access to the locks for cleaning and adjustment.

Barrels are taper-bored, a technique Lefever developed during the 1870s and used for all of his shotguns. Instead of beginning a few inches behind the muzzle, the choke is a function of the entire bore. Lefever believed that a straight, gradual taper from chamber to muzzle produced the most consistent, evenly distributed patterns.

Degrees and combinations of choke were entirely a matter of the customer's choice.

The Automatic Hammerless was produced in seven grades in 1885. Lowest were F, which sold for $75, and E at $100. They were plain, sturdy guns decorated with line engraving and stocked in plain American walnut. Grades D, C and B, which sold for $125, $150 and $200 respectively, were embellished with progressively more ornate engraving and stocked in English walnut. The two highest grades, A at $250 and AA at $300, were beautifully finished and engraved. Their barrels were of fine Damascus and stocks were of French walnut. Optional buttplates of buffalo horn or skeleton steel were available for all grades.

By the late 1880s Lefever was known throughout the gunmaking industry as Uncle Dan and his reputation as a designer was preeminent. For all the brilliance of his guns, Lefever Arms was in financial trouble and about 1890 Lefever sold controlling interest in the company to J. F. Durston and A. A. Howlett. None of Uncle Dan's earlier partnerships had

Lefever Arms Company catalogue, circa 1914-15. It lists several Durstons as company officers; the Durstons did not sell out to Ithaca Company until 1915.

198

lasted more than a few years, and this one was to be no exception.

Nevertheless, the Automatic Hammerless flourished. In 1881, the Ideal G Grade was added to the line, a utility gun that sold for $75. Two years later, the magnificent Optimus Grade was created and was featured at the Lefever exhibit at the World's Columbian Exposition in Chicago. The Optimus was a masterpiece of artistry, elaborately engraved with delicate scrollwork and inlaid with gold. Fitted with barrels of Whitworth fluid steel and stocks of the finest European walnut, the Optimus sold for $400. The Exposition judges awarded it First Premium and Diploma, given for excellence of design.

The harmony among Lefever and his partners didn't last long. By the turn of the century, Howlett was president of Lefever Arms, Durston was treasurer, and Uncle Dan was in charge of manufacturing. No doubt, it was the old story: executives on the one hand, urging economic restraint and production compromises; the old gun-maker on the other, refusing to give an inch where quality was the question. In 1901, Uncle Dan left Lefever Arms.

Ironically, he spent the last years of his life building guns that competed in the marketplace with other guns that already bore his name. In late 1901 or early 1902, Lefever and his five sons organized D. M. Lefever & Sons in Syracuse. The following year, the name was changed to D. M. Lefever, Sons & Company. Their catalogue stated: "not connected with Lefever Arms."

The Lefevers left Syracuse in 1903 and set up shop for a time in Defiance, Ohio. In 1905, the D. M. Lefever Company began operation in Bowling Green, Ohio. While Lefever Arms continued production of the Automatic Hammerless, Uncle Dan and his sons were building the New Lefever, a gun their catalogue described as "purely an invention of our own, with no foreign ideas attached."

Dan Lefever had no need for outside ideas, foreign or otherwise. Though radically different from his earlier guns, the New Lefever was to be yet another masterpiece.

The sleek sideplates of the Automatic Hammerless are gone. The bolting system features a Greener-type crosspin and a top lug, both of which engage a massive barrel extension at the top of the breech. The crosspin is of tempered cast steel and tapered to compensate for wear. There are other familiar Lefever features: the ball-and-socket hinge joint, the taper-bored barrels and the ingenious simplicity of the locks.

The famous Lefever cocking hook is there. A single milled-steel part, it performs five separate functions: it cocks the hammers, cocks the ejectors, acts as a primary extractor and an extractor-stop, and is a check-hook to cushion stress on the hinge joint.

Triggers are adjustable for weight of pull. Ejectors may be converted to extractors by turning a screw and instead of the traditional sliding button, the safety is a roller.

Lefever designed and patented three different single-trigger mechanisms. The last one, patented in 1903, is probably the safest, most efficient of all. Recoil blocks prevent the second barrel from firing until the trigger is released and returns to engage the second sear. Because there is virtually no friction, there is little likelihood of wear. The barrel selector is a cross-bolt through the stock cheeks and just above the trigger.

The 1905 catalogue listed seven grades of the New Lefever. All were built in 12, 16 and 20 gauges and, except for the 0 Excelsior, were standardly fitted with automatic ejectors. A single trigger could be ordered on any grade for an additional $15.

The D. M. Lefever line began with 0 Excelsior Grade at $60 and ranged upward through No. 9F, No. 8E, No. 7D, No. 6C, No. 5B and No. 4AA. The lower grades were stocked with English walnut and barreled in either Damascus, Krupp or Imperial steel. The No. 7D Grade was discontinued about 1904. Nos. 6, 5 and 4 were stocked with English or French walnut and barrels were of Damascus, Whitworth or Krupp steel. The No. 4AA Grade, which sold for $300, replaced an earlier No. 3 Optimus Grade that was discontinued about 1903. At the top of the line was the magnificent Uncle Dan Grade, a $400 festival of inlay and engraving.

> *Two gold medals were at stake, one for the best American gun and another for the best gun in the world. Dan Lefever won them both.*

The 1905 catalogue included another New Lefever, a single-barrel trap gun that sold for $38. Presumably made in higher grades as well, it was built in 12 gauge only and was offered with 26 to 32 inch barrels and weighing 6½ to 8 pounds.

The guns built at Bowling Green were initially stamped "D. M. Lefever, Sons & Co." and later "D. M. Lefever Arms Co." A great many of these shotguns, especially the higher grades, were not stamped with grade designations, which makes exact identification extremely difficult today.

Excellent gun that it was, the New Lefever was not particularly successful. Uncle Dan died in 1906, a patriarch of gun-making for 50 years. His last company died with him. Total production of D. M. Lefever guns numbered fewer than 2,000, making them the rarest of all American doubles.

Back in Syracuse, the Lefever Arms shotgun underwent little design change in the years following Uncle Dan's departure from the company. A number of minor revisions were made, presumably improvements planned by Lefever himself, for which the company held patent rights; but the gun was still the old Automatic Hammerless.

In 1913 there were 12 grades available in five gauges and selling from $37 to $1,000. Nearly half sold for less than $100. The utility gun, introduced about 1901, was called the DS Grade (for Durston Special) and was offered in 12, 16 and 20 gauges. An I Grade, nearly identical to the DS, was made for a few years and then discontinued. The H Grade had a bit of scroll engraving on the sideplates with stocks of English walnut and barrels of Best London Twist or Carman fluid steel. The $80 Grade F had Damascus or Premier Nitro Steel barrels chambered in 10, 12, 16 and 20 gauges.

The mid-range grades, E, D and C, were handsome guns priced at $100, $125 and $150. In all three grades the customer could order either English Damascus or Krupp fluid steel barrels and 10 through 20 gauges.

The high grades were works of art with beautiful English scroll engraving. All were barreled in English Damascus or Krupp steel, and the stocks were richly figured European walnut. The B Grade cost $200, A Grade $250, and AA Grade $300.

The $400 Optimus Grade was at the top side of the line. Select Circassian walnut stocks were matched with barrels of highest-grade hard English Damascus or Whitworth fluid pressed

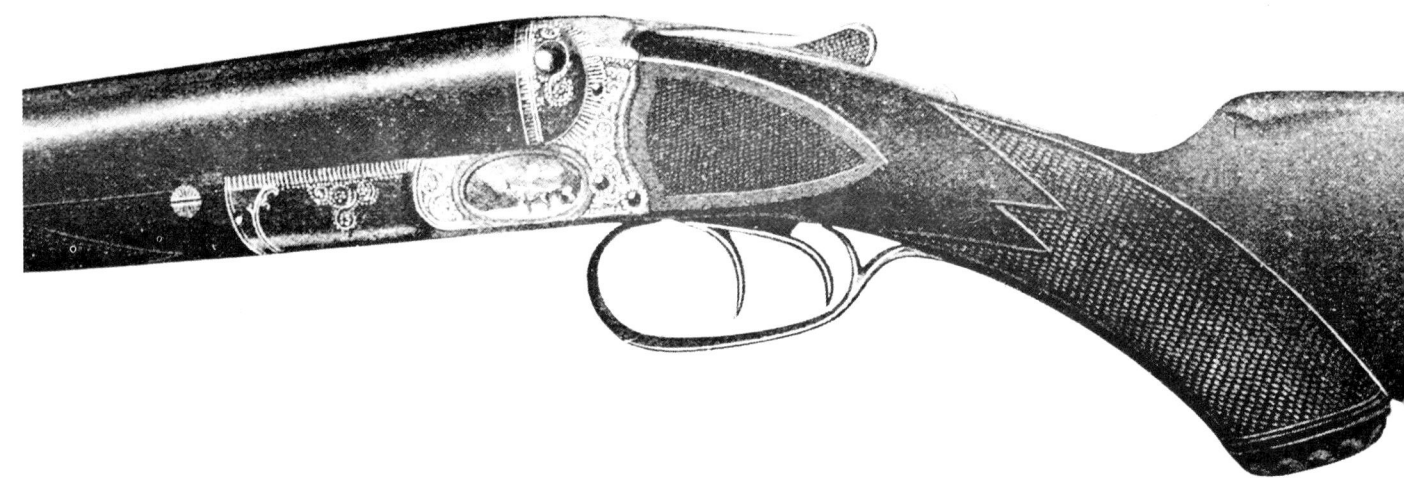

steel. Elaborate scroll engraving covered the frame and sideplates. Dogs and bird motifs were inlaid in gold on the plates, top latch, fore-end escutcheon, trigger guard and bottom of the frame. Thin gold bands highlighted the breech-end of the barrels and the safety button.

If that wasn't enough luxury, there was yet another grade. The Thousand Dollar Grade was completely custom-built and, at the time, the most expensive gun in America.

Automatic ejectors were standard in the four top grades and available in the rest for $15 above list prices. Guns of E Grade or higher could be had in 8 gauge for an extra $10. A single trigger

The D.M. Lefever line began with O Excelsior Grade at $60 and ranged upward through No. 9 F to No. 4 AA. This No. 8 E, a lower grade gun, was stocked with English walnut and barreled in either Damascus, Krupp or Imperial steel.

cost $25, its design typical of Lefever's innovation and meticulous attention to detail. The firing order was controlled by the safety slide. From the "safe" position, the button moved either forward or back; in the forward position, the right barrel fired first; moving the slide backward reversed the order.

Dan Lefever's guns always ran into financial ills. In the years just before World War I, competition among shotgun makers was especially fierce, and Lefever Arms sales began to slip. In 1915 the Durstons sold out to Ithaca Gun Company. The D. M. Lefever guns had been out of production for nearly 10 years. Now

Because they were built in such small numbers, the old Lefevers are hard to come by these days. A fair number of Nitro Specials turn up on the market and fetch modest prices, rarely more than $700 for the best of them and usually much less. The real gems are the Automatic Hammerless and the D.M. Lefevers. You'll look long and hard to find one, and when you do, it won't come cheap. An Uncle Dan Grade, in top-notch condition, will cost upwards of $8,000.

the great Automatic Hammerless was gone as well. In the 31 years it was in business, Lefever Arms produced only about 73,000 guns.

The Lefever name reappeared in 1921, this time as the Lefever Nitro Special which was an Ithaca-designed utility gun that sold for $29. Serial numbers began at 100,000.

Though it has little in common with its predecessors, aside from the Lefever name, the Nitro Special is a rugged gun. It was the first gun Ithaca designed specifically for modern, nitro-powder ammunition, and an advertising flyer dated February 15, 1922, claimed that the first lock was dry-fired more than 77,000 times and the first gun fired some 10,000 times without any malfunction.

The action is fitted with stout coil springs, and the stock is attached with a drawbolt that adds considerable strength at the wrist. The early Nitro Specials were made in 12, 16 and 20 gauges with .410-bore added in 1928.

Apparently the Nitro Special received an encouraging reception, for Ithaca brought out a single-barrel model in 1927. It was an unassuming hammerless gun that sold for $16. A trap gun, identical except for a ventilated rib, was also offered. Like the doubles, the single-barrel was chambered in 12, 16 and 20 gauges and in .410-bore. An A Grade double, only slightly more ornate than the basic Nitro Special, was added to the line in 1934.

As Ithaca tooled up for wartime production, sporting guns began to disappear. First to go was the Lefever single barrel in 1942; the Nitro Special went in 1948, along with all of Ithaca's double guns. The 70-year history of Uncle Dan's guns had come to a close.

Because they were built in such small numbers, the old Lefevers are hard to come by these days. A fair number of Nitro Specials turn up on the market and fetch modest prices, rarely more than $700 for the best of them and usually much less. The real gems are the Automatic Hammerless and the D. M. Lefevers. You'll look long and hard to find one, and when you do, it won't come cheap.

An 0 Excelsior Grade D. M. Lefever in reasonable condition costs about $750 and prices for the higher grades will go up in increments of $500 or more. An Uncle Dan Grade, in topnotch condition, will cost upwards of $8,000.

The Lefever Arms guns show an even greater price spread since there were more grades built. They range from about $600 for a DS Grade to about $2,000 for a C Grade; B Grade should bring about $2,500, A Grade about $3,500, and AA Grade about $4,500. You won't find many Optimus Grades selling for less than about $8,000, and a Thousand Dollar Grade will cost at least ten times its name.

That is serious money, any way you cut it. Nonetheless, the madness that has bloated Parker prices beyond reason hasn't yet caught up with Lefevers, making them a good investment both as collector pieces and companions for the field. The hardest part is finding one. Those who own one or two are hanging onto them and with good reason. There won't be any more guns like the ones that Uncle Dan built. ◆

SAVAGE 99
A Gun for the Ages

Perhaps the finest lever-action rifle design of all time, the Savage 99 is one of the world's oldest continuously manufactured guns and will still be going strong when many newer models have been laid to rest.

PETE LAURIE

In the latter stages of the 19th century the demand by hunters, soldiers and settlers for increased firepower produced the uniquely American lever action repeating rifles. During this period, the Henry, the Marlin, and the Winchester evolved. Perhaps the finest, most durable lever action design of all was that of A. W. Savage.

Apparently a jack of all trades, Savage followed a winding trail before becoming a pioneer in the development of sporting arms. He was born in 1857 in Jamaica, educated in England and the United States, and as a young man explored Australia in a covered wagon, eventually acquiring that country's largest cattle ranch, which he subsequently sold. He then returned to Jamaica to buy a coffee plantation. It was some years later while employed as a superintendent of the Street Railway in Utica, New York, that Savage turned his long interest in firearms and other mechanical devices to serious work on the development of small arms.

In 1892 he took out a patent on a revolutionary rotary magazine lever action, probably in cal. .30/40 Krag or .303. He attempted to sell the gun to the U.S. Army but it did not do well in tests and was not adopted. Savage modified his design over the next several years and in 1897 formed the Savage Arms Co. at Utica. In 1899 he introduced the gun that made him famous.

The Savage Model 1899 was a gun so far ahead of its time that it still is in production, the original design all but unchanged. The 1899 boasted several advantages over other lever actions. The enclosed, hammerless action prevented accidental discharge and was virtually immune to dirt and damage. The rotary magazine could cope safely with sharply pointed ammo that could self ignite in tubular magazines. The 1899 also had consistent balance not found in the easily damaged tubular magazines of other lever action rifles. Its side ejection had two distinct advantages over the

Many early Savage 1899s had fine engraving and checkering that equaled the very best. Photograph by Art Carter

top ejection system of many lever actions: it kept spent cartridges out of the shooter's line of sight and later made scope mounting possible. So strong was the 1899's lock up mechanism that it readily accommodated modern high power cartridges such as the .243 and the .308.

Jim Carmichel, gun editor of *Outdoor Life* recently wrote of the Model 1899: "Imagine, if you will, the impact of the Model 99 lever rifle on today's market if it had been introduced only last year. It would be hailed as a thoroughly modern, high-strength design using the best ideas in today's gun making technology... The 99 is one of the world's oldest continuously manufactured guns and will still be going strong when many newer models have been laid to rest. There is no better proof that a gun is great."

The Savage Model 1899 (in 1918 it assumed the simpler designation "Model 99", has been produced in more than two dozen models and better than a dozen calibers from the .22 Hi-power to the .375 Winchester. In New England and much of the Northeast, it became a standard brush gun alongside the Winchester Model 94. More than two million have been manufactured and the demand remains strong.

The first 99s could be purchased in six versions or "models" as they were called by Savage. The Model-A with a round barrel, the Model-B with an octagon barrel, and the Model-C with a half-round, half-octagon barrel were the mainstays. All three came with a 26-inch barrel, straight stock and curved steel butt plate. A short version of the Model-A was also offered with a 28-inch barrel and full military stock. The Model-F was a carbine with a saddle ring.

During the past 80 years, modifications to the original 99 have been minor but noteworthy. About 1904, around serial number 90,000, the cocking indicator was changed from a raised block on the bolt to a pin on the upper tang. At the same time the rear surface of the bolt was changed from a flat surface to a rounded surface. Today, Savage does not recommend firing a 99 with a serial number under 90,000 due to the possibility of the receiver cracking, although 99s with the flat bolt probably still exist and have been in continuous use without damage.

In 1920 the original versions of the 99 were replaced with seven new models which incorporated minor modifications of the basic mechnical design. In subsequent decades a confusing array of various models were produced

for a number of years and then discontinued. Many of these later models bore the same model designations as earlier versions. D. P. Murray's book, *The Ninety-Nine,* is a good source to sort out some of this confusion. Also in 1920 a takedown version was introduced that later could be purchased in a boxed combination with a .410 shotgun barrel. The .410 barrel screwed into the receiver and was attached to the stock with a special screw. In the .410 mode, the 99 operated as a single shot since the rotary magazine could not accommodate the bulky shotshell. The combination was designed to be a camp gun, so that hunters could take birds and other small game while hunting deer, bear, elk, or moose. In the days when big game hunting often involved a long trek on foot or horseback into a remote camp, the combination eliminated the need to tote a smoothbore just to shoot meat for the pot. The boxed combination was discontinued in 1934 and the .410 barrel discontinued in 1940.

In 1965 a removable clip magazine was offered because many states had passed laws prohibiting loaded weapons in vehicles. The clip or box magazine allowed the 99 to be loaded and unloaded more quickly and safely.

The early 99s came with all manner of cosmetic refinements that boosted the cost upward from the standard price of $20-$25. The expansive, flat sides of the 99's receiver were ideal for engraving and several types were offered by the factory in grades listed from A to G. The simple scroll work in the A grade cost an additional $5 in 1900 while the elaborate hunting scenes of the F and G grades ran the purchase price to more than $200. In some grades scroll work graced the lever/trigger guard and all manner of nickel, silver, and gold plating were offered.

By the 1920s the factory had apparently abandoned the costly custom engraving and in 1926 introduced the Deluxe Engraved Model 99-K. In a recent article Murray offered this description of the 99-K:

"Although the quality of finish and engraving can't compare to the special order 1899s made earlier, the 99-K retains the quiet dignity of a firearm suitable for the sporting use of the most discriminating shooter. The fancy select walnut stocks were hand checkered with a larger pattern than the 99-G, and even the flat buttstock side panels carried a nicely checkered area. Leafy scroll engraved panels cover each side of the

The one millionth Savage 99 features beautiful wood and the Savage emblem gold inlaid into the right side of the receiver. Photograph courtesy of Guns & Ammo.

receiver ring, barrel, and area near the rear of the breech bolt have simpler designs.

"The action of the 99-K was carefully fitted and hand stoned to assure a smooth trigger pull and tight action lockup. The takedown barrel was made in two lengths — 22-inch for the .22 H.P., 250-3000, .30/30, and .303 Savage; 24-inch chambered for the .300 Savage. Special sights consisted of a two-leaf rear barrel sight and Lyman folding tang peep sight. The front bead/blade sight was mounted in the modern ramp. The usual corrugated steel shotgun type buttplate finishes up this nice piece. The 99-K was made through 1940, serial range 290,000-398,000."

The longest running models of the 99 were the R and RS (identical to the R except for the sights). Introduced in 1930 these models had a heavy semi-beavertail forearm with a blunt tip and were fitted with sling swivels. They were chambered for fairly heavy cartridges. Both were discontinued in 1960 after almost 30 years of production.

The Model EG was introduced in 1933 and later modified with a checkered pistol grip and a triangle design on either side of the forearm. The EG was dropped in 1960 but a great many are still in circulation.

Stocks varied from straight to pistol grip and later had Monte Carlo cheek pieces. In the early years a "perch belly" style was quite popular. A variety of fore-ends were offered from a barrel length musket to a simple metal-banded brush or saddle gun, but a graceful schnabel tip was typical of most 99 fore ends. Rocky Mountain type buckhorn sights were standard but a variety of adjustable sights could be added. By the 1950s most models were drilled and tapped for scope mounts.

After World War I bolt action rifles grew in popularity with American sportsmen due in part to the experience soldiers gained with military rifles such as the Springfield and Mauser. This shift in interest coincided with the marketing of several excellent bolt actions most notably with Winchester Model 70. Accuracy and pressure packed power or high velocity took precedence over rapid second and third shots.

With more leisure time, more money, and better transportation, the ultimate hunting ex-experience evolved from a rocking chair buck in the hollow below the barn to a bull elk in

Close-up of the one millionth Savage 99, dated March 22, 1960 and presented to the National Rifle Association.

the Rockies or a grizzly in Alaska. Other hunters discovered the fun of experimenting with the hot little wildcats that could kill a woodchuck at 300 yards. For these types of shooting the lever actions generally could not touch the stronger, more accurate bolt actions.

For many Easterners white-tailed deer hunting remained a favorite sport. Since deer present a large target at usually short ranges and are not difficult to kill, such old standbys as the .30/30, .35 Rem, and .300 Savage retained their popularity. The durable, reliable 99 became a tradition. However, in the South where heavy brush and short ranges many times resemble hunting conditions in the Northeast, the 99 has never been as popular. This is due in part to the widespread use of buckshot by generations of Southern deer hunters.

The first 99s were chambered for the Savage .30 and .303. The .30/30 cartridge became available in 1900. By about 1903, around serial number 30,000, the 99 was chambered for .25/35, .32/40 and .38/55.

The Model 99 helped write the history of American ammo development with the .250/3000 introduced in 1913. The first standard factory load to exceed 3,000 feet per second, this little cartridge, now known as the .250 Savage, it is still in production and 99s are still chambered for it.

The .30/06, long a standard with American gunners, did not adapt well to lever actions where short bolts and quick lever throw are of prime importance. So in 1920 the .300 Savage, a shortened version of the .30/06 with almost the same ballistics, was developed specifically for the 99. The .300 Savage later was modified into the 7.62 NATO round and re-introduced for sporting rifles as the .308 Winchester in 1952. The ability of the 99 to cope with the extreme pressures generated by the .308 and later by the .243 (another offshoot of the .308) has ensured its continued popularity with American hunters.

Long-time shooting editor of *Pennsylvania Game News*, Don Lewis, calls the 99's longevity its greatest testimonial. A gunsmith who has drilled and tapped hundreds of 99s, Lewis also claims to have fired "as many 99s as anyone."

"The Savage 99 has always been popular in this part of the country," Lewis notes, "It is a favorite of left-handed shooters, especially since the introduction of the shotgun style safety on the top of the receiver.

"Although the current craze for magnums has swung some hunters away from lever actions, a great many older hunters still carry Savage 99s, Winchester 94s, and Marlin 336s," he says.

Although Lewis is a bolt action aficionado, he admits that a hunter's choice of action is usually based solely on personal considerations.

"For most North American big game, the hunter himself plays the key role in the success of the hunt, not the rifle or the cartridge. Such old standbys as the .300 Savage, .250 Savage, and the .30/30 would not still be around after all these years were they not perfectly adequate for whitetails and most other big game," he adds.

Calling the Savage 99 a standard big game rifle with a long reputation for reliability and endurance, Lewis foresees no decline in the future demand for this gun.

The Savage 99 is a classic, a gun of another age that remains modern and functional. The deluxe, dressed-up versions manufactured early in this century today have no equivalents at any price. Even if you own a plain vanilla, battered and scarred from many years in the deer woods, you have a weapon your grandchildren will one day use. The 99 will be with us for a long, long time.♦

CHARLIE PARKER'S
Shotgun

With only $70 and a blind horse, Charles Parker started a business that would eventually produce America's most famous shotgun, a gun so famous, in fact, that even Russia's Czar Nicholas II wanted one.

MICHAEL McINTOSH

In 1832 shotguns were the furthest thing from Charles Parker's mind. He wanted to build coffee mills. He was 23 years old, going into business for himself, and his operating capital consisted of $70 and a blind horse.

Born at Cheshire, Connecticut, in 1809, Parker had come to Meriden in 1828 following an apprenticeship at a Southington, Connecticut, button factory to work for Patrick Lewis. Lewis made coffee mills, and young Parker soon saw there was money in the hardware business. With the few dollars he's saved, and with the blind horse hitched to a pole sweep as a source of power, he built a prosperous life.

By the time the Civil War began, Parker was one of the largest hardware manufacturers in New England. His catalogue listed literally

---

The A-1 Special 12-gauge Parker, directly above, has deep relief engraving. A prime example of this gun, priced at $895 when it was made in 1915, would be worth $50,000 today. The top gun is a 1920 VH grade Parker. A step above the Trojan grade, it is valued at $2,500. Photograph by Richard W. Smith. A-1 Special 12-gauge Parker and VH grade Parker, courtesy of Chaddick's Ltd., Terrell, Texas.

211

hundreds of items — everything from pumps and hinges to silverware, door knockers and waffle irons — distributed and sold all over the world. The war prompted Parker's first interest in gun-making, and by the time Lee and Grant met at Appomattox, the company had built almost 17,000 rifles in aid of the Union cause: 15,000 Model 1861 Springfields made on government contract and the rest breech-loading repeaters in .50 and .55 calibers, designed by the Parker engineers.

According to Peter Johnson, the first serious Parker historian, Charles Parker had for years nurtured a yen for building a fine shotgun. That he waited 30-odd years to do so befits the conservative nature of the man. Parker was no innovator; all of the hardware he built was homely, tried and true, notable more for quality than novelty. English makers had established the double's basic design more than a century before, and although there was plenty of room for improvement, Parker knew that a high-quality sporting gun would find a ready market.

What he could not have known was that he was about to create the most famous shotgun ever built in America.

Design work began in 1865 and though the first prototype may have been completed that year, the Parker was no overnight phenomenon. Like most great guns, it grew out of what had come before — in this case, patents that Parker bought from Smith & Wesson and design features patented by William H. Miller in 1866. By the time the evolution of the old-style Parker was complete, it incorporated features designed and patented by such men as John Stokes, Joseph Dane, Francis Dangerfield, Charles King and Dan Lefever.

The first Parker gun was available for sale in 1868. It was a 14-gauge hammer gun with 29-inch barrels. In most respects, it was a typical shotgun of its time, but one feature, at least, was truly unique. The bolting system — a sliding bolt that engages a notch in the barrel lump — is operated by a lifter-type plunger in the floor plate just ahead of the trigger guard. It's an ungainly looking affair and, to one accustomed to the now-conventional top latch, awkward to operate, but it lacks nothing in strength or reliability.

Over the first few years, Parkers were built in 8, 10, 11, 12, 14, 16 and 20-gauges. Though the guns were breech-loaders, fully self-contained shotshells were not yet in general use, and the Parkers fired a transitional type of shell comprising a brass case, powder, wadding and shot but no primer. The guns were fitted with nipples for standard percussion caps which, when struck by the hammers, sent a spark through a flash channel in the standing breech to ignite the powder through an opening in the shell base.

By 1872 Parkers were offered in six grades, distinguished by the type of locks and the quality of the barrels. Bar-action locks, with the mainspring and sear placed in front of the hammer, were used in the three higher grades; the lower grades were fitted with locks of the back-action type — mainspring and sear behind the hammer. The barrels were various grades of Damascus twist, imported as blanks from Belgium. Low-grade guns were barreled in plain twist, higher grades in Damascus of progressively better figure and density.

The second phase of Parker evolution began in 1874, when Charles A. King was hired. Previously employed at Smith & Wesson, King was a brilliant designer, and he brought the Parker gun truly into the modern age.

The first thing to go was the old lifter action.

The single most critical factor in the design of a break-action gun is the means by which the barrels are fastened to the frame. By King's time, gunmakers had tried virtually every mechanical principle from wedges and pins to screws, cams and clips. Some worked; some didn't, but no definitive system had emerged — nor has it yet. There are two or three basic systems in use today that all seem to work equally well. The one that King finally chose for the Parker was the sliding underbolt, standard even then, but he added a refinement of his own that makes the Parker bolting system one of the world's best.

Instead of bearing against the barrel lump itself, the locking bolt engages a small steel plate pinned to the rear of the lump. The bearing surfaces of both plate and bolt are angled at a pitch of 12½ degrees, providing a constantly tight fit without undue friction, even as the steel wears through use. The steel plate can be replaced to return the fit to factory-new tightness.

To help distribute stresses and to act as secondary locks, King designed a pair of interlocking hooks, one fastened to the barrel lump and one inside the water-table slot. With the hooks engaged, the action would remain closed even if the gun were fired with the main locking bolt retracted. To prevent any lateral play between barrels and frame, King used the doll's-head rib extension designed by Dan Lefever, which fits into a milled notch at the top of the standing breech.

The new bolting system went into production in 1882. Except for a minor revision of the

Old illustration of Meriden, Connecticut, home of the Parker shotgun. Charlie Parker arrived here in 1828 and began working for Patrick Lewis, a manufacturer of coffee mills. After four years with Lewis, Parker started his own hardware business.

barrel-lump plate in 1910, it remained for as long as the Parker gun was made.

In 1889, the first hammerless Parkers, designed by Charles King, were produced. Always conservative, however, Parker continued to produce hammer guns until about 1915 and built them on special order as late as 1920.

By 1899, Parkers were available in 23 grades, 14 of them hammer-type and nine hammerless. Hammer guns were offered in A.A. Pigeon, A, B, C, D, E, F, G, H, I, R, S, T and U grades; the A.A. Pigeon sold for $400 and the U Grade for $50. Hammerless guns also were graded by letter designations, with an "H" added to indicate a hammerless action. When automatic ejectors were developed in 1902, an "E" was added to the grade names of those guns that had them. The "H" remained a part of the Parker grading system as long as the guns were built — and long after hammer guns were discontinued.

The Trojan Grade, the nearest thing to an economy gun that Parker ever built, was introduced in 1915 at a price of $27.50.

With the steady improvement of self-contained shotshells, small-bore guns became truly practical around the turn of the century. In 1905, Parker built the first American breech-loading guns in 28-gauge. Chambered for a 2½-inch shell that held ⅝ ounce of shot and 1¾ drams of black powder, 28-gauges were built in all grades, though the majority of them were high-grade guns. Only those Parkers factory-chambered for the 3½-inch Magnum 10-gauge and for the .410-bore are rarer.

The Trojan Grade, the nearest thing to an economy gun that Parker ever built, was introduced in 1915 at a price of $27.50. The Trojan frame has less graceful contours than the higher grades; no doubt manufacturing costs were reduced by eliminating certain cosmetic milling operations. Early Trojans had the doll's-head rib extension, but later ones, built after the mid-1920s, did not. Strictly a production gun, the Trojan was made only in 12, 16 and 20-gauges and was never made with ejectors or beavertail fore-end or as a trap or skeet gun. The fore-end latch is a tension-hook rather than the mechanical latch that Charles King designed in 1880 and that was used on all other Parker grades as long as the guns were built.

Though the hammerless Parkers proved a great success, there were still problems. King's action is highly complex — 18 parts in all — and no doubt was expensive to machine and difficult to fit. James P. Hayes, who had worked with King in developing the Parker ejector system, began redesigning the action about 1910 and over the next few years succeeded in reducing the number of component parts to only four. The new action, which went into production about 1917, marked the final phase in the Parker's mechanical evolution.

The new-style Parkers were available in ten grades: A-1 Special, AAHE, AHE, BHE, CHE, DHE, GHE, PHE, VHE and Trojan. The Trojan sold for $43.50 and the A-1 Special for $600. Automatic ejectors were standard in the top five grades and, except for the Trojan, optional in the rest.

In response to the growing popularity of trap shooting, Parker introduced a single-barrel trap gun in 1917. It was built in 12-gauge only with a 30, 32, or 34-inch barrel, fitted with a vent rib. There were five grades, corresponding to the double-gun grades, with an added "S." SC Grade was the lowest, followed by SB, SA,

SAA and SA-1 Special. Prices ranged from $150 to $550.

No other major American gun-maker ever built shotguns in as many different gauges as Parker. The 10-bore was predominant when Parkers were first built, and it remained the most popular gauge until the turn of the century, when improved ammunition led to the supremacy of the 12-gauge. Ten-gauge Parkers were available as long as the guns were built, though few were made after World War I. Eight-bore Parkers were available almost from the beginning and remained until about 1915. Before 1917, G and P grade guns could be had in 14-gauge, and there were even a few 24-gauge Trojans built sometime after World War I. Parkers chambered in .410-bore were introduced in 1927.

In order to maintain standards of weight and handling quality, Parker built guns on seven different frames, each scaled in size and weight to the shell for which it was chambered and to barrel length. The lightest frame, No. 000, was for the .410-bore, No. 00 for 28-gauge; both have troughs milled out of the water table to further reduce weight. No. 0 frames were used for 28, 20 and 16-gauges, No. 1 for heavy 20 and standard-weight 16, No. 1½ for lightweight 12 and heavy 16, No. 2 for standard 12 and lightweight 10, and the massive No. 3 frame was for the heavy 12, standard 10 and the monstrous 8-gauge.

By the beginning of the Great War, more than 160,000 Parker guns had been built and sold, and Parker's reputation reached halfway around the world. Though Major Hugh Bertie Campbell Pollard, the great English arms historian, later remarked that it was an adequate shotgun "for farmers' use," Parkers were well thought of in Europe, especially in France. The greatest tribute, however, came from Nicholas II, Czar of Russia, who about 1917 directed the Russian Consul in New York to place an order for a Parker gun. Nicholas was an enthusiastic hunter, and his gun room was replete with specimens of Europe's finest sporting arms. The Parker was to be his only American gun.

The Czar's Parker was a 12-gauge A-1 Special, its stock trimmed in gold, the imperial Romanov eagle inlaid in gold on the trigger plate — a gun fitted and finished with all the skill the Parker craftsmen could muster. Nicholas never saw it. Revolution plunged Russia into a nightmare of blood just after the order arrived at Parker's, and in July 1918 Nicholas and his family were slaughtered in the cellars of an old house in Ekaterinburg. The Parker eventually was sold in this country, but its fate has remained a mystery for nearly 70 years.

Though the A-1 Special was the highest of the production grades, as tastefully ornate and meticulously finished as any shotgun needs to be, at least two Parkers were even more lavishly treated. As the story goes, Parker sought to commemorate its 200,000th shotgun by building what the factory described as "the finest example of the gun makers' art produced by an American gun maker." It was to be called *Invincible*.

Production of Parker's single trigger mechanism began in 1922.

The first Invincible was built early in the 1920s, a 12-gauge fitted with a straight-grip Circassian walnut stock. Except for the barrels, every inch of steel was engraved in fine English scroll, and game birds sculpted in gold were inlaid on the frame. The serial number, 200000, was inlaid in gold on the trigger-guard tang.

Apparently, the gun was put on display at various places around the country. After a well-publicized appearance at Kennedy Brothers Sporting Goods in Minneapolis, it was shipped west, possibly bound for Kansas City or Omaha, and was never seen again. How it disappeared and where it is now has become almost the stuff of legends. All that remains is a photograph that appeared in the 1926 and 1930 Parker catalogues.

The little plant that turned out 6,000 guns per year during the 1920s built only about 500 in 1932 and 1933 combined. The most famous gun in America had become a liability.

Legend has it that a second Invincible was built at the same time as No. 200000, a companion gun with serial number 200001. There is no solid evidence that it ever existed.

So much myth and rumor has accrued to the Parker Invincibles that it's difficult to winnow out the few, sketchy grains of truth. There was, however, at least one other Invincible, delivered from the factory on Friday, September 13, 1929, to Mr. A. C. Middleton of Moorestown, New Jersey. It was a lovely gun, a 16-gauge with 26-inch barrels, choked improved cylinder and modified and chambered for the old-style 2 9/16-inch shells. The half-pistol grip stock was intricately carved and checkered, the engraving and inlay virtually identical to the first Invincible. It was fitted with double triggers — the front one hinged in the English style — ejectors and splinter fore-end. The serial number is 230329.

The stock market collapse in October 1929 must have distracted Middleton from his new Parker. By the time he died some years later, the gun was all but forgotten. His widow died in the early 1950s, and a couple of years after that the family house was sold. The new owner found, in an upstairs closet, an unfired Invincible Grade Parker in its factory leather case, the hang-tag still attached to the trigger guard. The gun ended up in a private collection and then, in the late 1960s, in a gunshop at Ridgefield, Connecticut, about 60 miles from the factory where it was built. It sold again in 1972.

The stock market crash and subsequent Great Depression proved to be as tough on the Parker company as it was on those who owned the guns. Much of the previous decade had been spent developing refinements, and the Parker of 1930 was as refined as the Parker ever was. The single trigger was first offered in 1922, the beavertail fore-end in 1923, and a vent rib for the double guns in 1926. Damacus barrels, the beautiful, treacherous holdovers from black-powder days, were discontinued in 1925. The January 1930 catalogue shows the doubles available in nine grades (the P Grade was discontinued sometime after 1922) and the single trap gun in five. But as the American economy crumbled and breadlines lengthened, even the $55 Trojan Grade was a luxury that few could afford.

Parker still owned its old hardware business, which greatly improved the company's prospects for survival, but the gun works fell into deeper and deeper doldrums. The little plant that had turned out 6,000 guns per year during the 1920s built only about 500 in 1932 and 1933 combined. The most famous gun in America had become a liability. Finally, it was offered for sale.

Remington Arms bought the Parker gun on June 1, 1934. With the infusion of financial support, production slowly revived. In 1937, Remington moved the Parker operation, along with the machinery and a great many of the employees, to its factory at Ilion, New York.

Though no design changes were made, the Remington-built Parkers lack something of the fine quality of earlier days. The mechanical integrity remained, but checkering and engraving, especially, show less care and skill. Remington-made guns can be identified by the word *Parker* engraved on the bottom of the frame just ahead of the trigger plate.

As it did for so many of the fine old American double guns, World War II spelled the end of the Parker. Remington converted most of its manufacturing resources to military production in 1942, and the few Parkers assembled during the war and after, no doubt were made from parts already machined. When sporting guns once again went into production late in 1945, the old Parker machinery remained silent.

Serial number 242385 is generally believed to be the last Parker built. I've had the privilege of holding No. 242387 — a G Grade .410, assembled in 1947, that now resides in a private collection. It has never been fired.

In intrinsic quality, there really is little to distinguish a Parker gun from other fine American doubles. The old Foxes and Lefevers and L. C. Smiths were built and finished with every bit as much skill and attention to detail.

Yet Parker generally is the most highly regarded, the most sought-after of all. Parkers fetch the highest prices, figures that surpassed the limits of reason years ago.

But price and value have little to do with one another, and if Parkers are overpriced, they certainly are not over-valued. There is scarcely any other item of human manufacture that combines such practical and aesthetic values as a fine gun, and its worth far exceeds the sum of its parts. Its value rests in the workings of the human spirit — in the creative process by which it was conceived and in the beauty and function it communicates.

The Parker gun is therefore an artifact, a relic of a quieter, more graceful past, a human achievement undiluted by time. A Parker tells us that beauty of form and function have a place in a day-to-day world where beauty is too often obscured. For that reason alone, it has value beyond measure. ♦

Besides being the founder of the Parker Company, Charlie Parker was also the first mayor of Meriden.

Classic QUARTET

Four classic .22 rifles — Winchester Model 90, Remington Model 12, Browning Automatic, and Marlin Model 39 — continue to be a vital part of Americana — from plinking fields and target ranges to collectors' cabinets.

RICK HACKER

The .22 repeating rifle — how unjustly maligned it has often become simply because it is so very much a part of our lifestyles. The ultimate "first" gun for a youngster, protector of the farmer's henhouse, procurer of meat for many a family during the Depression and well beyond, and camp gun for the hunter. For the knowledgeable collector, the .22 rifle is more than just a part of Americana. It embodies an important link to firearms history. Besides being a symphony of trombone-action, lever-action and semi-automatic harmony in wood and steel, the .22 repeater holds many hidden variations that take it out of the realm of the ordinary and place it in the upper stratosphere of the high-grade rarities, a unique phenomenon for so common a firearm.

The .22 round itself, the most inexpensive and universally available cartridge made in history, accounts for much of the .22's popularity. In fact, the venerable .22 rimfire has been in existence longer than any other self-contained metallic cartridge, having first made its appearance in 1845. Since that time, it has been chambered in everything from pip-squeak-powered single shot Flobert-styled parlor pistols to rapid-firing miniature Gatling guns. Yet its real claim to fame in this century was due, in large measure, to four classic .22 repeating rifles. Paradoxically, the rifles themselves would probably never have rated very high on the popularity polls were it not for the fact that they were chambered for the .22 round. Our classic quartet shares two other common characteristics: each gun was extremely well

Left to right: Winchester Model 90 courtesy of David Moore; Remington Model 12 courtesy of Chadick's Ltd.; Browning .22 Automatic courtesy of Chadick's Ltd.; Marlin Model 39 courtesy of Marlin Firearms. Photograph by Art Carter

made and each had a take-down feature that enabled it to be easily transported in a bedroll, suitcase, or backpack (a feature that somehow seems more advantageous to yesterday's shooter than today's).

By far the most popular and eagerly sought of the take-down .22 rifles is the Winchester Model 1890, a slide-action, octagon barreled, exposed-hammer pump gun that first appeared in sporting goods stores across the country just prior to Christmas 1890. Winchester obviously had high hopes for their new .22 repeater and chose to spotlight the Model 90 prominently on the back cover of their 1891 catalog, noting that it sold for just $16. Interestingly, this was the same price of Winchester's 1866 Sporting Rifle and was just $3.50 less than their Model 1886 carbine featured in the same catalog.

Initially, the Model 90 was offered in a solid-frame version; the familiar take-down model was introduced one year later, after some 15,500 solid-frame guns were produced. The early solid-frame models are extremely desirable among collectors, and in pristine condition they can bring figures in the four-digit range. All Model 90s came with a one-piece brass cleaning rod, an accessory that is rarely found in any of the guns encountered today.

Originally, the Model 90 was offered in models chambered in .22 Short, .22 Long and .22 W.R.F. (Winchester Rim Fire), cartridges that had been developed expressly for this gun. The .22 Short proved to be the most popular cartridge until 1919, when the .22 Long Rifle chambering came along and quickly unseated all the rest. It was far more accurate and more effective for hunting both small and medium-sized game, and in those pre-conservation years, the .22 was often used on animals as large as deer. However, the .22 Short continued to prove popular in Model 90s that were relegated to the shooting galleries in countless small town carnivals and state fairs across the country. No matter how many thousands of rounds were put through it, the little pump just kept on shooting and rarely jammed. This perfection of design can be credited to the Model 90's inventor, John M. Browning, the firearms genius who is responsible for three of the four classic .22s discussed in this article.

Browning was 32 years old when he first conceived of the Winchester Model 90, but he never intended the pump gun for the diminutive .22. Originally, the rifle was designed for much heavier calibers, per a request from the Winchester executives for a big bore pump rifle that could compete with the popular Colt slide action .44-40. The young gun designer met the challenge admirably, but by the time his slide-gun was perfected, Winchester realized they already held a marketing lead in big bore repeaters with their Models 73, 76 and 86 lever-actions. Rather than pass up the excellent action of the Model 90, however, Winchester decided to scale down the massive pop-up sliding breechblock and sturdy yet simple inner workings of the Browning design to a .22 size. Browning's concept adapted

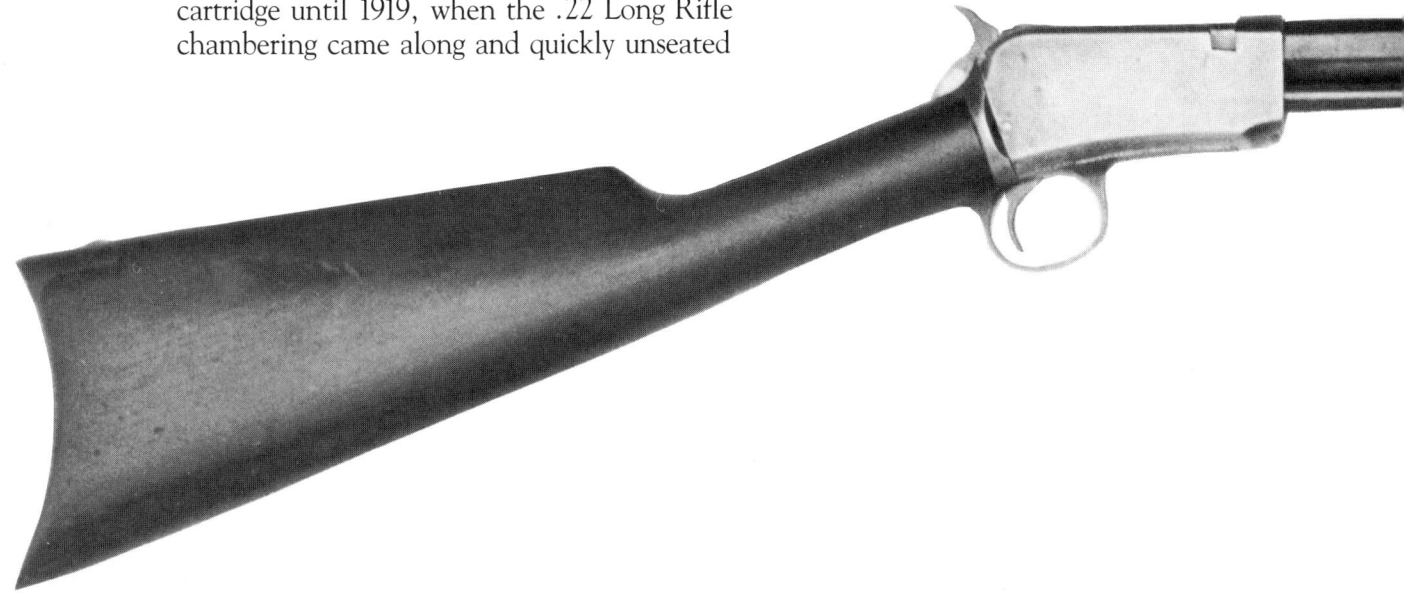

to the task easily and the modification enabled the Model 90 to outlast generations of shooters. Many of these rifles are still in use today.

Browning designed the Model 90 for safety. It could not be fired unless the action was completely closed and locked. In addition, the massive protecting bolt could not be opened while the hammer was cocked unless the shooter manually and purposely depressed the firing pin. The choice of using highly corrosive ammunition, however, was destructive to many Model 90 rifles. Remember, these were the days of black powder loads and mercuric priming compounds, and with hard-hitting cartridges such as the .44-40 and .45-70 high on the popularity polls, few people took the .22 seriously. Winchester's own catalogs stated, "It will not be found necessary to clean the action of the gun..." when referring to the Model 90, although the company did advise shooters to swab out the bore. But few people read beyond the initial sentences, and this misplaced "shoot it and forget it" attitude is the reason why so many Model 90s are found with their barrels completely rusted out. Fortunately, replacement barrels for the Model 90 (as well as the other three classic .22s) can be obtained from Numrich Arms, West Hurley, New York 12491.

It was partly due to the corrosion problem that Winchester instituted a promotion in the 1920s that urged owners to send their rifles back to the factory to be fitted with new barrels made of stainless steel. The offer created a variation of the standard model which is extremely rare today, for evidently not many shooters wanted to take the time and trouble to have their Model 90 fitted with a different barrel. As a result, a Model 90 factory-fitted with Winchester stainless steel barrels (which are marked as such) commands a premium when it can be found and authenticated.

For the first eight months of production, the receivers, buttplates, hammers and trigger guards of the Model 1890 were casehardened. However, from the August 1891 production run until the last rifle was assembled in 1941, the standard finish was blue. Winchester was not anxious to offer options on their new .22, no matter how popular it might become. Early catalogs stated: "Guns will be furnished only with 24-inch octagon barrels, plain triggers, and straight grip rifle stocks. We are not prepared to furnish longer barrels, set triggers or pistol grip stocks." However, customers were used to ordering any extra feature they wanted and could afford, and eventually Winchester bowed to demand and provided a few special-order guns with engraved receivers, fancy wood, pistol grip stocks, hand checkering and special sights. Casehardened receivers were also available as a special order item, a little-known fact that can often confuse collectors examining Model 90s with serial numbers higher than the 15,500 range.

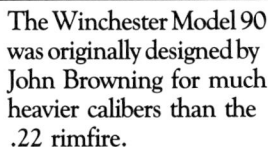

The Winchester Model 90 was originally designed by John Browning for much heavier calibers than the .22 rimfire.

One drawback to the Model 90 was that its chamberings were not interchangeable (i.e., a .22 Long would shoot that cartridge and no other), a situation eventually remedied by the Model 90's alter ego, the lower-priced Model 06.

By 1921, the price of the Model 90 had risen to $31.50, while the 06, which could handle all three .22 cartridges, was selling for $28.75. The Model 90 was finally dropped from the Winchester line in 1932 and replaced by the Model 62 pump, which boasted the interchangeable .22 feature. This was not the end of the beloved Model of 1890, however, for just prior to World War II the last readily-available cache of Model 90 parts was assembled into rifles, many of them fitted with round, rather than octagon, barrels. This final production run brought the 51-year span of the Model 90 to about 849,000 guns.

In spite of the advent of newer, more practical designs, the Model 90 continues to endure. In 1981, while on safari in Zimbabwe, I stopped at a little sporting goods store in Victoria and saw a well-worn Model 90 for sale in the gun racks; it had only recently been traded in by a rancher who wanted a more "modern semi-automatic." And in the 1960s, while living in Arizona, I purchased a vintage Model 90 chambered for the .22 Short for only $15. The slim little pump served me well for many years as my desert jackrabbit gun, and it continues to do well on California cottontails. Aside from the rare variations (where the law of demand is the only true guideline for price), most Model 90 "shooters" can be bought for about $150; mint specimens, exhibiting all of their original finish, bring prices that start around $275 to $350. Of course, early solid-frame Model 90s command a premium, as do those with factory "extras" and engraving.

If the Model 90 can be called the most popular rifle in our classic quartet, then the Remington Model 12 is certainly the most respected. First introduced in 1909, the sleek little pump had a number of features that made it a favorite of hundreds of thousands of shooters during its 27-year lifespan (it was not discontinued, but replaced by the Model 121, which remained in production from 1936 until 1954). For one thing, the Model 12 had a solid-topped breech, which protected the action from rain, dirt, and snow. A sliding safety, located just behind the trigger on the guard, locked the trigger. The absence of an exposed hammer lessened the opportunity for an accidental discharge. Finally, and of immense benefit during those early, relatively competition-free years, the Model 12 could shoot .22 Short, Long, and Long Rifle cartridges interchangeably. In addition, a separate .22 Short-only Gallery Special could be had for no extra charge, with a special 1-in-24 inch rifling twist as compared to the 1-in-16 inch twist found on the Standard rifle. The Model 12 weighed 4½ pounds (a full pound less than the Model 90) and, like the other three rifles in this article, sported a take-down feature for ease of cleaning from the breech.

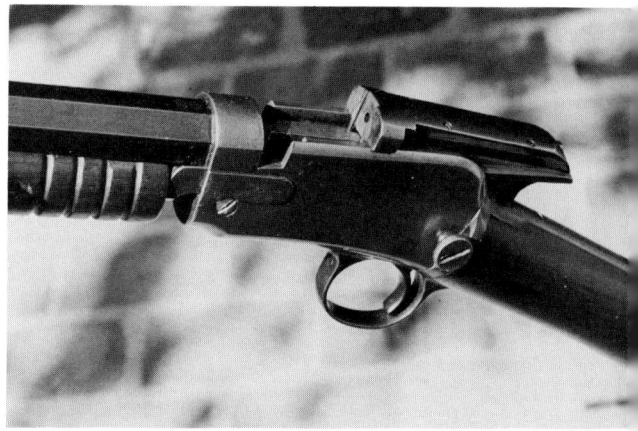

The 90's sliding action with its heavier-than-necessary bolt adapted easily to the modification.

The Model 12 was a favorite among gallery owners; besides being practically jam-proof, it was extremely safe. With a round fully chambered, the slide locked and the solid-topped receiver kept any escaping gases away from the shooter's eyes. An overly-large ejection port spat empty cases out to the side of the gun and also enabled a shooter to insert a single round into the breech. Small wonder that the Model 12 was a favorite of Annie Oakley, who thought enough of the rifle to pose with it for promotional posters and magazine covers.

The Standard (Field) Grade was introduced in 1912 and sold for $12.65. The rifle featured a round barrel, straight-grip stock and a flat rubber buttplate. The Gallery Special held 15 .22 Short cartridges in its tubular magazine and came with a pistol grip stock, crescent buttplate and octagon barrel. The Gallery Special could also be ordered with a choice of rear sights that included "special fine," "regular," or "special coarse" (undoubtedly on most of the Model 12 Gallery Specials I fired in my youth). It is interesting to note that while the Standard Grade Model 12 could be ordered in .22 Short chambering only with no increase to the $12.65 price, the Gallery Special sold for $16.

The Target Grade, also listed at $16, handled all three .22 cartridges and featured a straight grip stock. With its crescent buttplate and octagonal barrel, the Target Grade was probably the handsomest version of the Model 12 available at an affordable price. Both the Target and Gallery versions could be special ordered with an extra-long tubular magazine for an additional five dollars, thereby increasing the rifle's capacity of .22 Shorts from 15 to 25 rounds. Normally the gun held 12 Long and 11 Long Rifle cartridges, the same as the Model 90.

The Peerless Grade sported a pistol grip checkered stock of selected walnut and a superbly engraved game scene that covered 100 percent of the receiver and the trigger guard. Available for just the .22 Short or for all three versions of the .22 cartridge, this octagon-barreled model took a tremendous jump in price to $45 (a substantial amount of money in 1912 for any rifle, not just a .22 pump-action). As a result, very few Peerless models are encountered today.

Another top-line version of the Model 12 is the Expert Grade. Listed in the 1912 catalog for $60, it featured scroll engraving over the receiver, trigger guard and portions of the barrel. All internal parts of the action were hand polished. The finely-figured walnut stock was hand-checkered and available either with a straight grip or a pistol grip (full grip or half grip was also at the customer's option).

Finally, there was the Premier Grade, so-called because it sported the same full-coverage, quality engraving found on Remington's Premier Grade auto-loading shotgun of that era. All other features were the same as on the Expert Grade Model 12. The price for the Premier was $75, and I have never seen one for sale. It is uncertain how many of these engraver's masterpieces were produced, although a total of 831,000 of all Model 12s were made prior to the introduction of the Model 121.

In 1921 Remington produced a No. 12 N.R.A. Target Grade chambered for just the .22 Long Rifle cartridge. This Model 12 featured a leather sling strap that was only fastened to the magazine tube and not the buttstock, thus clearly marking the rifle for target shooting. The N.R.A. model also came equipped with a tang peep sight and a "windage combination aperture globe" hooded front sight. The rifle sold for $43.89; in that same year the cost for the Standard Grade had risen to $28.48 and the

Gallery and Target grades to $31.95. Interestingly, the price for the "extra long" magazine tube had dropped to $3.50. A variation of the Model 12 was the Remington Special Grade (Model 12 CS), which was the same gun as the Target Grade, except that it was chambered only for the .22 W.R.F. cartridge (which Remington called the Remington Special).

Today, Standard and Gallery Special Model 12s can often be found for sale in the $175 to $225 price range. Target Grades are slightly higher, depending on condition. The engraved models, however, are in a class by themselves and are rarely encountered, except in advanced collections. Yet interestingly, many Model 12s are still in use today, especially the Gallery rifles, which rarely show any signs of wearing out.

Another .22 rifle that simply refuses to die is the Browning Automatic, a somewhat antique-looking semi-automatic that was first introduced to American shooters in 1956. However, far from being a "modern" gun, it actually appeared on the shooting scene 32 years earlier, wearing a different external disguise: in 1924 it was known as the Remington Model 243.

John Browning had originally patented his .22 automatic in 1912 (he received the final patent in 1914), and he had a few hundred of his blow-back operated rifles manufactured in the well-known Belgium factory of Fabrique Naitonale D'Armes De Guerre (F.N.). These early Browning .22 automatics were marked, "Browning Arms Co. — Ogden, Utah" and bore the date of the last patent, Jan. 6, 1914. The rifles were sold exclusively through the Browning retail store in Ogden, Utah, from 1914 until the outbreak of World War I. Obviously, these pioneer semi-automatics are quite rare. The rifle did not appear on the market again until 1924, when Remington brought it out as their now-famous Model 24; in 1935 the name was changed to the Model 241 and the sleek little repeater stayed in the line until it was discontinued in 1951.

In 1956 Browning Firearms brought out their own version of their founder's invention, dubbed simply the Browning .22 Automatic. Although most of the internal mechanism was based upon the same principle as the old Model 24, the gun was completely new externally. The stock was fuller and had a deeper, more dramatically sculpted pistol grip; the thick, enlarged forearm filled the supporting hand and made the Browning comfortable to hold while still retaining its slim look. Ironically, the Browning .22 looks as if it had come out of the early 1900s, yet it sports one of the most modern and jam-proof designs of any semi-automatic rifle on the market today.

From its introduction in 1956 until 1973 the Browning .22 Automatic was made in Belgium, and these models always command a 30 percent to 50 percent increase in prices at gun shows. They can often be found in their original boxes. The .22 semi-automatic has been made in Japan since 1974; I have one in my collection and consider it a superb shooter.

The current Brownings are tapped for a scope sight (the early rifles did not have this feature)

This Grade III Browning features a hand-checkered forearm and extensive engraving. It sells for $815 and holds considerable promise for the collector.

and are chambered for either the .22 Long Rifle (magazine capacity of 11 rounds) or for the .22 Short (holding 16 rounds). The Browning has a unique tubular magazine that withdraws from the buttstock. Loading is done through a scalloped hole in the right-hand side of the stock. Collectors should be interested in the fact that the Browning .22 Automatic is the only "classic .22" that can still be purchased in a choice of grades.

Grade I is the most common and features a blued finish with a moderate amount of scrollwork on both sides of the receiver. This model retails for $267.95 and is often discounted by some of the larger retail sporting goods chains — a good value if you plan to shoot it. The Grade II sells for $380 and sports a satin finish chrome-plated receiver with a hand-engraved scene of two squirrels on the right of the frame and two prairie dogs on the left side. The top of the receiver depicts a duck in flight (presumably for those areas in which waterfowl hunting with a rifle is legal!).

The Grade III Browning, I believe, holds the most promise for the investor/collector since fewer of these guns are made each year and the engraving is extensive (the receiver is fully engraved and the engraving extends onto the trigger guard as well). The left side of the receiver depicts a setter flushing two pheasants in a forest, while the right side shows a brace of ducks being flushed by a Lab of dubious ancestry. The top of the receiver shows the dog catching the birds. On some of the early guns, the artists actually signed their names and occasionally, the engravers were given carte blanche to create their own scenes. These variations, while esoteric to many collectors, make the Browning unique and valuable. Currently, the Grade III sells for $815.

Today, of course, customized Brownings are rare indeed, although a few special-order and presentation-grade guns are known to exist. In addition, it is a little known but welcomed fact that engraved initials and other inlays can be special-ordered on the Browning .22. As the owner of one Browning rifle with my name on the receiver, I can say that the service is first-rate!

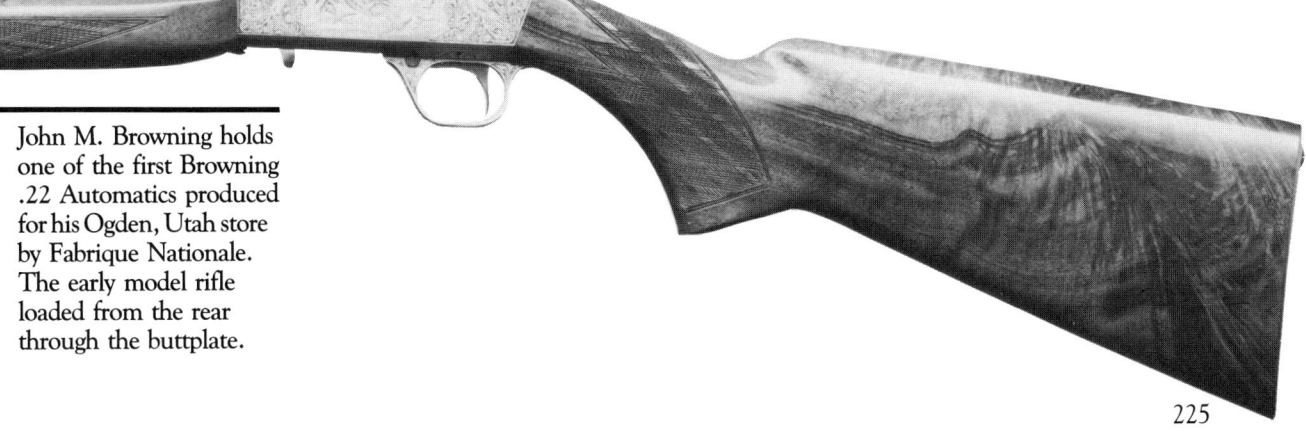

John M. Browning holds one of the first Browning .22 Automatics produced for his Ogden, Utah store by Fabrique Nationale. The early model rifle loaded from the rear through the buttplate.

A unique feature of the Browning Automatic is the knurled knob at the breech end of the barrel, an adjustment device that keeps the rifle from loosening after years of take-down activity. From 1956 until 1963, buttstocks of the automatic were attached to the receivers via a 5½-inch bolt running through the pistol grip. Since then the stock has been attached with a round-slotted nut on the magazine tube channel. If a replacement stock is needed, it is important for the owner to differentiate between the two types, which can also be used to help date the gun.

John Browning was 58 years old when he invented his .22 automatic. He had already achieved immortality in the world of firearms with Winchesters 86, 92, 94, and 95. Yet it is his little .22 that continues to live on, new-and-in-the-box. There must be a message in there somewhere.

Another classic that is still produced is the Marlin Model 39 lever-action .22 rifle. It has the distinction of having the longest continual production run by the original manufacturer of any rifle in the world. It is also the first .22 repeater that was made to fire Shorts, Longs, and Long Rifles interchangeably. The year of its introduction was 1891, and it was called, logically enough, the Marlin Model 91. Until that time, the only .22 lever action in existence was the Winchester 73, and that toggle-link action, while famous for the .44-40, could hardly be called a practical solution for the 29- to 40-grained .22 slug.

Originally, the Marlin .22 loaded from a sideport like most of the lever-actions of its day, but one year after its introduction it became the Model 92 and sported the now-familiar loading tube underneath its barrel. In 1897 another significant improvement was made with the addition of the take-down feature, a system the current Model 39 retains to this day.

Of his new Model 97, John Marlin stated, "It costs more to make and will cost you more than the other .22 rifles, but it's the best .22 caliber repeater in the world if you are willing to pay for it."

Evidently John Marlin was not the only one to think his .22 lever-action was "the best." Tom Mix, Captain A. H. Hardy, T. K. Lee, and numerous other crack marksmen found favor with the Model 97. But the most famous was Annie Oakley; she fired hundreds of thousands of rounds with a Marlin 97 during her numerous tours with Buffalo Bill's Wild West shows. With the fast-shooting .22, she broke glass balls, snuffed out candle flames, and split playing cards. One of Ms. Oakley's most publicized feats was the time she fired 25 shots in 27 seconds at a playing card that was 36 feet away; she placed every shot through the center of the card. The rifle she used was a Marlin 97.

After World War I, Marlin changed the name of its now-famous .22 (by then sales had surpassed the quarter-million mark; 125,000 of which were Model 97s) to the Model 39 and in 1937, after implementing a few minor internal and external design changes, the gun became the 39A. Marlin's Micro-Groove rifling was added in 1954 and in 1957 the company brought out their carbine, known as the Mountie, which featured a 20-inch barrel and straight grip stock. The rifle holds 26 Shorts, 21 Longs and 19 Long Rifle cartridges, while the carbine holds 21, 16, and 15 respectively.

The Model 39 has been used for various presentations to celebrities and politicians. A pair of unique guns with which I am personally acquainted are the factory nickleplated, smoothbore Marlin 39Ms used from 1953 to

1961 by actress Gail Davis, who played Annie Oakley in the popular television series. Gail did all of her own trick shooting when she went on publicity tours throughout the U.S. and Canada, literally duplicating the shooting feats of the real Annie Oakley. For her act, she had one Marlin 39 with a carved stock, keeping the second gun, nearly identical but without the carving on the stock, for a backup. It is a tribute to the ruggedness of the Marlin's action that the backup gun was only used once in over eight years and countless thousands of rounds. Such one-of-a-kind guns are not often encountered.

More readily available to collectors on the used gun market is the Marlin 39 Century Ltd., a special commemorative put out by Marlin in 1970 to honor its 100th anniversary. The Article II carbine and rifle are still found in gun shows and two years ago were selling for $300 to $400 apiece, about $100 over the regular list price of the standard Model 39.

Like the Winchester Model 90, the Marlin 39's action was originally designed for a more powerful cartridge. Its locking system first appeared on the Marlin Model 1881, a lever-action that was chambered for the .45-70. Clearly, the Marlin 39 has a fine frontier pedigree behind it, although it is the earlier Models 91, 92, and 97 that are often sought by collectors (even though these rifles rarely bring more than the retail price of the new guns). Rifles in pristine condition or with factory engraving, however, are quite rare and can command a premium.

The Marlin was always considered a "workhorse" .22 and, as a result, saw rugged use. (Then) Captain Charles Askins, writing in his 1935 booklet, "Shooting Facts," put out by the Outdoor Life Recreation Library, wrote of the Marlin 39, "This is a well-balanced, handy rifle, well-stocked and it is doubtful if a better small game rifle is to be had."

While there is not as much collector value in the Marlin 39 as might currently be found in some of the earlier versions such as the 1891, 92, and 97, it is only because the little lever gun is still in production. Since 1921, more than a million Model 29s have been made, yet if production ceased, they would immediately assume a status all their own. It was the very same way with the Model 90 and the Model 12; the Browning, due to its Belgian ancestry, is already assuming collectible notoriety.

Ironically, when not being coveted in a collector's cabinet, these classic .22s still manage to find their way out to the target ranges and the plinking fields. Whether in the hands of a young girl or boy just learning the rudiments of firearms training, or being held by an experienced sportsman who appreciates the craftsmanship and heritage these collectibles represent, our four take-down smallbores occupy a unique niche in firearms shooting and collecting history. ♦